Springer

Tokyo
Berlin
Heidelberg
New York
Barcelona
Budapest
Hong Kong
London
Milan
Paris
Santa Clara
Singapore

S. Nishikawa
R. Schoen (Eds.)

Lectures on
Geometric
Variational
Problems

With 15 Figures

 Springer

Seiki Nishikawa
Professor, Mathematical Institute,
Tohoku University,
Sendai, 980-77 Japan

Richard Schoen
Professor, Department of Mathematics,
Stanford University,
Stanford, CA 94305-2125, USA

ISBN-13: 978-4-431-70152-1   e-ISBN-13: 978-4-431-68402-2
DOI: 10.1007/978-4-431-68402-2

Preface

In this volume are collected notes of lectures delivered at the First International Research Institute of the Mathematical Society of Japan. This conference, held at Tohoku University in July 1993, was devoted to geometry and global analysis. Subsequent to the conference, in answer to popular demand from the participants, it was decided to publish the notes of the survey lectures. Written by the lecturers themselves, all experts in their respective fields, these notes are here presented in a single volume. It is hoped that they will provide a vivid account of the current research, from the introductory level up to and including the most recent results, and will indicate the direction to be taken by future research.

This compilation begins with Jean-Pierre Bourguignon's notes entitled "An Introduction to Geometric Variational Problems," illustrating the general framework of the field with many examples and providing the reader with a broad view of the current research. Following this, Kenji Fukaya's notes on "Geometry of Gauge Fields" are concerned with gauge theory and its applications to low-dimensional topology, without delving too deeply into technical detail. Special emphasis is placed on explaining the ideas of infinite dimensional geometry that, in the literature, are often hidden behind rigorous formulations or technical arguments. The third set of notes, by Leon Simon, entitled "Theorems on the Regularity and Singularity of Minimal Surfaces and Harmonic Maps", is intended to introduce the analytic aspects of the study of regularity properties and singularities of minimal surfaces and harmonic maps. These notes are based in many places on the results and techniques developed recently by the author himself, which in various ways unify and simplify the arguments in the literature.

It is the hope of the editors that these surveys in geometric variational problems will prove to be a powerful stimulus for further research into these fields.

March 1995

Seiki Nishikawa
Richard Schoen

Table of Contents

An Introduction to Geometric Variational Problems

Jean Pierre Bourguignon

Centre de Mathématiques
U.R.A. 169 du C.N.R.S.
Ecole Polytechnique
F-91128 PALAISEAU Cedex
France

I have been asked to present, in a series of four lectures, an *introduction to Geometric Variational Problems*. This theme has its place in the History of Mathematics since its development has often been strongly tied to the birth of many of the topics presently covered under the heading Geometry. Nevertheless, it remains extremely active, and its zone of influence has reached many new areas of Mathematics in the last half of this century. A proof of that vitality is given by the many lectures at this Institute quoted in this survey.

In view of the great variety of topics to cover, presenting Geometric Variational Problems as a whole is a formidable task. I was therefore forced to make very personal choices. These notes, which complete on a number of places the lectures[2] actually given, are, as a result, partial. I nevertheless hope that they can still serve as an introduction to the subject as a whole.

The material of this survey has been divided into three chapters :

- Chapter I : *A General Scheme*
- Chapter II : *A Review of Geometric Variational Problems*
- Chapter III : *Symmetry Considerations, Topological Constraints and Interactions with Physics*

---

[1] The author gratefully acknowledges the support of the Japan Society for the Promotion of Science during his visit to Tôhoku University at the invitation of Professor Shigetoshi Bando in the summer 1993 when this report was completed.

[2] The author thanks the organizing committee of the First Research Institute of the Mathematical Society of Japan entitled *"Geometry and Global Analysis"*, in particular its chairman Professor Seiki Nishikawa, for the excellent facilities provided for the conference, and the auditors of the lectures for their patience and positive attention.

Chapter I

The General Setting

The purpose of this chapter is to present the general setting in which Geometric Variational Problems develop. It would be of little meaning were it not accompanied by the numerous examples listed in Chapter II. Hence, it is advisable to take it seriously only after one has read Chapter II.

The first section describes the general framework of a Geometric Variational Problem. A second section is devoted to presenting the various points of view that one can adopt in dealing with these problems and to the various uses one makes of solutions of Variational Problems in Geometry, and more generally in Mathematics.

1. A General Framework

1.1.   We always work in the context of differentiable manifolds and bundles. (A good introduction is [140].) We most of the time assume that the data are C^∞ but, from time to time, we will be forced to work with less regular objects. In fact, being able to do so has lead to great progress in the field in the last quarter of this century.

GENERAL SET-UP 1.2.— *A Geometric Variational Problem consists of a real-valued functional $\phi : \mathcal{O} \longrightarrow \mathbf{R}$, where $\mathcal{O} = \Gamma^{-1}(c_0)$, the space of geometric objects, is a subset of a space $\mathcal{S}$ of sections of a bundle $\pi : E \longrightarrow M$ with $\Gamma : \mathcal{S} \longrightarrow \mathcal{C}$ a constraint map and $c_0 \in \mathcal{C}$, a point in the constraint set. The base manifold M is often endowed with a Riemannian metric[3] that we generally denote by g.*

Most of the time, $\mathcal{O}$ is defined in such a way that it retains more structure than just being a set. The most one can hope for is of course that $\mathcal{O}$ be a manifold, and $\phi : \mathcal{O} \longrightarrow \mathbf{R}$ be differentiable. This is for instance the case when $\mathcal{C}$ is a manifold and Γ a submersion. In this context, solving a variational problem consists of determining all critical points of ϕ, and analyzing their geometric properties.

---

[3] since taking norms defined by a scalar product is the most natural way of reducing geometric quantities to scalars. Therefore, a basic knowledge of Riemannian geometry is assumed. Good introductions to this discipline are [19], [36] and [68].

In less structured cases, one is left with studying whether ϕ achieves its infimum (or its supremum), and, if yes, how the minima (resp. maxima) behave. Even, in the smooth case, minimizers (resp. maximizers) of ϕ, i.e., points in $\mathcal{O}$ at which ϕ achieves its minimum (resp. maximum), exhibit often some special properties, and are worthy of further investigations.

> *For the rest of this chapter, we will assume that the space $\mathcal{O}$ is a manifold and that ϕ is as differentiable as necessary so that one can speak of the differential of ϕ and of its higher derivatives.*

1.3.   Since critical points of a differentiable function are characterized by the vanishing of its differential, one of the key steps in setting up a Variational Problem is *to establish the First Variational Formula of the functional ϕ*, i.e., to compute for each $u \in \mathcal{O}$ its linearization $d\phi(u)$.

The equation $d\phi(u) = 0$ is often referred to as the *Euler-Lagrange equation* of the problem[4]. This is often a *system of Partial Differential Equations*. Its type (linear or non-linear, elliptic, parabolic or hyperbolic) depends on the variational problem as we explain later.

When ϕ is expressed as an integral involving an element $u \in \mathcal{O}$ and some of its derivatives, taking the differential of ϕ is often referred to as computing the *variational derivative* of the integral because of the special expression to be used in order to take properly into account the presence of derivatives of u in the integrand.

1.4.   As is well known, the local behaviour of a function, like ϕ, near one of its critical points u_0 is not determined by first order data (precisely because $d\phi(u_0) = 0$). At u_0, it is then possible to define *intrinsically* the *Second Variation* of ϕ or its *Hessian*, by setting, for $X, Y \in T_{u_0}\mathcal{O}$, $\mathrm{Hess}_{u_0}\phi(X,Y) = \partial_X(\partial_{\tilde{Y}}\phi)(u_0)$ where $\tilde{Y}$ denotes an extension of the tangent vector Y to a vector field defined in a neighbourhood of u_0. Notice that $\mathrm{Hess}_{u_0}\phi$ is a symmetric bilinear form on the space $T_{u_0}\mathcal{O}$. (Recall that we assume ϕ to be as differentiable as necessary so that the Schwarz lemma can be applied.)

The nature of $\mathrm{Hess}_{u_0}\phi$ is directly related to the *stability* of the critical point u_0 of ϕ, hence *the importance of computing the Second Variation Formula of the functional for any Geometric Variational Problem.* Indeed :

---

[4] to remind us that Leonhard Euler established the first version of this Calculus in an analytic setting, and that later Joseph Louis de Lagrange [83] generalized it to a more general (today one would say non-linear) context. At that time, it was not yet appropriate to use the word *"points"* to speak about elements of a function space. Volterra [139] is probably the first to have thought of the problems in the Calculus of Variations as infinite dimensional versions of the determination of critical points of a function defined on a configuration space.

– if u_0 is a *maximum* of ϕ, then $\mathrm{Hess}_{u_0}\phi$ is *negative semi-definite* ;
– if u_0 is a *minimum* of ϕ, then $\mathrm{Hess}_{u_0}\phi$ is *positive semi-definite* ;
– if $\mathrm{Hess}_{u_0}\phi$ is *non-degenerate*, then the converse is true ;
– if $\mathrm{Hess}_{u_0}\phi$ is *indefinite*, then u_0 is a *saddle point*[5] ;

The kernel of Hess_{u_0} is usually called the *nullity* of ϕ a t u_0, and· its dimension the *nullity index*. Another invariant of $\mathrm{Hess}_{u_0}\phi$ is also of great importance, in particular in connection with Morse theory, namely the dimension of the largest subspace of $T_{u_0}\mathcal{O}$ on which $\mathrm{Hess}_{u_0}\phi$ restricts to a negative definite form, called the *index* of ϕ at u_0.

2. A Rough Classification of Geometric Variational Problems

2.1.   It is worth emphasizing here again that a Variational Problem depends on *all* the data required to define it : $\mathcal{S}$, $\mathcal{C}$ and Γ, and of course ϕ. It is convenient to divide Variational Problems into families according to which of these objects is taken of primary importance.

2.2.   A possible criterion is to take into account the *type of objects* on which the Variational Problem is defined, i.e., the nature of the space $\mathcal{S}$.

Here, we list a few typical examples :

– *functions* on manifolds, certainly the simplest objects ; the theories based on them are referred to as *scalar field theories* by physicists ;

– *tensor fields* on manifolds : the most classical examples correspond to *vector fields*, *exterior differential forms* (with a special emphasis on 1-*forms*, 2-*forms*, and *volume elements*), *Riemannian metrics*, *Lorentzian metrics* (because, as we shall recall later in Chapter III, they lie at the heart of General Relativity) ;

– *spinor fields*[6], which are only defined on manifolds endowed with bilinear forms that we take most of the time positive definite, i.e., Riemannian metrics ;

– *maps*, which, in certain cases, can be thought of as parametrising (sometimes with some additional properties) intrinsically defined geometric objects (e.g., *curves* if the source space is 1-dimensional, *surfaces* if it is 2-dimensional, *submanifolds* if it is higher dimensional).

---

[5] Notice that one can make sense of such points only by using the differentiable structure whereas local maxima and local minima can be identified by purely topological means.

[6] Notice though that, unlike physicists who attach great importance to the value of the spin of a particle, hence consider many irreducible representations of the spinor group, mathematicians tend to focus their attention only on the fundamental irreducible spinor representations. This limitation does not seem justified.

2.3.   Another way of discriminating among Variational Problems is to focus one's attention on the functional ϕ itself. The key point then is to determine its *non linearity*, a question which also involves understanding the constraint map Γ

The simplest case is met when $\mathcal{O}$ is a *vector space* (or an open set in a vector space) and ϕ is *quadratic*. Then, the Euler-Lagrange equations are *linear*, hence significantly simpler to study.

This leaves us with the other cases, i.e., when ϕ is more complicated than quadratic and/or $\mathcal{O}$ more sophisticated than being merely an open subset of a vector space. The *type of non-linearity* of ϕ and the *curved nature* of $\mathcal{O}$ of course determine the *non-linearities encountered in the Euler-Lagrange equations*. We postpone a more thorough (and hopefully less general) discussion of this point until we have explicit examples at hand.

When the Euler-Lagrange equations are elliptic Partial Differential Equations, some critical exponents connected to the conformal behaviours of ϕ and $\mathcal{O}$ play a crucial role. They draw a line between different types of behaviour for the solutions.

2.4.   In discussing a Geometric Variational Problem, another key feature is also to give a full description of its *symmetries*, i.e., transformations of the space $\mathcal{O}$ which leave ϕ invariant. Are also of importance maps of $\mathcal{S}$ which transform ϕ in some very specific fashion.

We postpone the discussion of this (very important) point until Chapter III, where we will derive some of the consequences on the Variational Problem that one can draw from symmetries, or generalizations of them.

3. Different Points of View on Geometric Variational Problems and Their Uses

A) The diversity of Geometric Variational Problems

3.1.   Geometric Variational Problems can be (and are) approached in several quite different ways.

One of my aims in writing these notes is to try to bridge a possible gap between classical considerations on this domain of Geometry and more modern viewpoints. This is not at all to say that classical domains are not any more active. Quite to the contrary, some of the most fundamental problems, which have a rich history, are still either unsolved or only partially solved, hence the fact that they remain important. Just to name one, the *isoperimetric problem* (cf. [103] for a survey) is still intensively studied, and fundamental questions concerning the regularity of its solutions in not so general spaces are still unanswered.

Nevertheless, it can prove useful, along with the systematic presentation given in Chapter II, to organize Geometric Variational Problems in families, a task we now pursue.

3.2.   One may be interested only in giving a *variational characterization* of some geometric objects. The knowledge of the objects preexists to this characterization which is just a way of singling them out among other similar objects. Needless to say, in this case, the functional ϕ, and most of the time also the constraint space C and the constraint map Γ have to be manufactured, and many choices are possible, not all of equivalent interest.

Most often, even if the variational approach to a geometric object starts in this way, the possibility of defining the object as a critical point of some functional brings some more information on it, e.g., one can attach to the point the index (and nullity index) of the functional. This may suggest to look for other related geometric objects.

This was (more or less) the traditional way in which geometers considered Variational Problems. In some sense, they relied on geometric constructions to get at the objects, but were interested in knowing more on their geometric nature[7].

3.3.   Another point of view puts all the *emphasis on the functional* ϕ in case it has a legitimacy of its own. This is typical in problems motivated by Physics and Mechanics, as we shall discuss in Chapter III.

Then, solving the Euler-Lagrange equation attached to the functional is the heart of the matter. Since most of these equations are non-linear, needless to say, we are far from being able to give a general answer.

On the other hand, because of the variational origin of the problem, it has been repeatedly verified that these equations exhibit especially nice behaviours. The *regularity theory* of these equations is particularly rich. Indeed, if one considers *weak* solutions of these equations, i.e., solutions in the distributional sense, they quite often turn out to be classical solutions, i.e., genuine differentiable solutions. This is very often true for solutions which *minimize* or *maximize* a sufficiently coercive functional.

The so-called *direct method in the Calculus of Variations*, which has recently been applied successfully to a number of problems, works along these lines. For a sequence of elements $(u_j)_{j \in \mathbf{N}}$ of $\mathcal{O}$ on which the functional ϕ approaches its infimum (assumed to be finite), it can often be shown that one can extract a *weakly convergent subsequence* because ϕ is strong enough to

---

[7]   A typical instance of that can be traced back in the study of lines in classical Euclidean geometry which are not considered as shortest paths in Euclid's *Elements*, but rather as building blocks of the Geometry, an attitude usually called a *synthetic* approach. Nevertheless, considering them as shortest paths was certainly instrumental in Gauß' thoughts about non-Euclidean geometries.

be proper for the weak topology. One then needs to show that the limit (an object belonging to an appropriate completion of $\mathcal{O}$) is a classical solution, the so-called *regularity step* in the Geometric Variational Problem

3.4.   Notice though that the traditional approach to the Calculus of Variations lies somewhat in between these two extreme attitudes. Indeed, on the one hand, this Calculus was introduced to determine objects by their behaviour under *"variations"*. This assumes that the objects are indeed known. On the other hand, because of the great success of Variational Principles to lay the foundations of Mechanics and Physics, and, as a consequence, of the great variety of situations in which solutions of Variational Problems were sought for, there has been a huge amount of work devoted to developing the second point of view.

In some sense, one can consider *Global Analysis*, the more modern expression which has replaced *"Calculus of Variations in the Large"* (the english translation of the german *"Variationsrechnung im Großen"*) introduced earlier in this century, as a synthesis of these two view points[8]. Thanks to major recent progress in the theory of Partial Differential Equations, first for linear ones via functional analytical methods, then for non-linear ones via the extension of these constructions to objects defined on manifolds and bundles, many more equations of variational type can be solved today. The major new interplay has been with topological constraints. Because this point is not so often publicized in the context of Geometric Variational Problems, we decided to devote a separate section to it.

3.5.   The development of *Morse theory* (cf. [96] for a historical account and [94] for a superb presentation) has played an important role in establishing Global Analysis as a discipline in itself. It continues to attract a lot of attention with new points of view, cf. [146]. Indeed, this theory allows a topological description of a space, say $\mathcal{O}$, on which a function ϕ with non-degenerate critical points with finite indices is defined, by giving a cellular decomposition of $\mathcal{O}$ obtained by attaching a k-dimensional handle in a well-defined way at each critical point of index k of ϕ. The basic ingredients of this theory are a *deformation lemma* asserting that two level sets $\phi^{-1}(\alpha)$ and $\phi^{-1}(\beta)$ can be deformed onto one another if the interval $[\alpha, \beta]$ does not contain any critical value of ϕ, and the so-called *Morse lemma* showing that in appropriate coordinates a function is quadratic near a non-degenerate critical point.

This possibility of investigating the global "shape" of a space via the detailed knowledge of critical points of functions defined on it gives another reason to study functionals on spaces of geometric objects, i.e, to consider Geometric

---

[8] In this respect, the survey [51] and the book [104] have played an important role in establishing the subject.

Variational Problems. A good example[9] of such an instance can be found in [12].

Of course, another way of using Morse theory in the context of Geometric Variational Problems is to use in the opposite direction the relationship it establishes between the topology of the space and the critical points of functions defined on it, i.e., by *obtaining the existence of critical points of the functional ϕ from a knowledge of the global topology of the space $\mathcal{O}$*. For this to work, one needs to know a priori that the critical points of ϕ are non-degenerate, and that ϕ is proper in some sense. A possible expression of this "properness" is the so-called *condition (C) of Palais-Smale* [107]. The study of functionals, which do not satisfy this condition and for which one can still put a modified version of Morse theory to work, has recently attracted a lot of attention.

Because of the non-degeneracy condition that Morse theory requires on critical points of ϕ, other tools have been developed to deal with functionals with possible degeneracies, e.g., *Lusternik-Schnirelmann theory* (cf. [105]). This theory is also based on a deformation lemma, and on the fact that the number of critical points exceeds the number of contractible sets necessary to cover the space.

B) The Role of Singularities

3.6.   So far, we only alluded to the fact that, although we will mainly be working with smooth objects, we may have to consider less regular objects. This has recurrently been one of the very fruitful themes of study in the Calculus of Variations (see a more complete discussion of this point in Chapter II, and the lectures by L. Simon at this conference). But such a presentation is too restrictive for at least three concurrent reasons ;

> – *in order to be sure that a given solution of a Geometric Variational Problem does not exihibit singularities, it is important to study how they can occur*, hence to study them in their own right. The occurence of singularities can preclude the use of global arguments because some basic formulas relating the local geometry of the space and global invariants require that the data be smooth, e.g., the possible failing of the Gauß-Bonnet theorem if infinite concentration of curvature occurs. Having definite information on the structure of the singular set, e.g., a bound on its codimension, can prove decisive in applying some geometric arguments ;
>
> – *there are* indeed *interesting solutions of Geometric Variational Problems which present singularities*, the most obvious case being that of

---

[9] One may wonder why functionals inspired by physical considerations give rise to functions having such a good behaviour from the point of view of Morse theory.

soap films spanned by some wires, e.g., the frame of a cube. Their actual realization in real life assures us that they are stable critical points of a functional of which the area is a good approximation ;

in some cases (sometimes because of the physical origin of the problem, but this may happen too on purely mathematical grounds), it has even become important to *consider problems with prescribed singularities*. In that case, the question remains to control the singularities.

C) The Various Uses of Geometric Variational Problems

3.7. Here, we list some of the main uses of solutions of Geometric Variational Problems in Geometry :

- *to select representatives* in spaces which are a priori interesting via one of their quotients (e.g., geodesics among homotopic curves, sub-manifolds in a given homology class, harmonic forms among forms in a given cohomology class, etc.) ; in that case proving that the functional ϕ has a single critical point is of course crucial, a fact which is often achieved by proving that ϕ is convex ;

- *to distinguish canonical objects* among general ones ; this is a way of giving alternate characterizations of objects which have already called the attention of geometers ;

- *to suggest estimates* by proving that a given functional takes its values in a given interval ; it may indeed turn out that calculating its critical values is simpler than calculating a generic one ;

- *to create new objects* ; a typical case is given by spaces of critical points of some Geometric Variational Problems ; one most often refers to these spaces as *moduli spaces* ; they come equipped with a lot of extra structures which make their study interesting and somewhat easier.

Chapter II

A Review of Geometric Variational Problems

This chapter presents a review of Geometric Variational Problems, hence has a long list of sections. Each section begins with the presentation of the most classical example under this heading using the general set-up introduced in Chapter I. Then, we move on to give some less classical versions of Variational Problems and some geometric consequences that can be drawn using this type of Variational Problems.

1. The Eigenvalue Problem for the Laplacian on Functions

A) The Classical Eigenvalue Problem using Rayleigh-Ritz Quotients

1.1.   Let $\mathcal{F}$ denote the space of smooth functions on a compact Riemannian manifold (M, g). The space of geometric objects is the space $\mathcal{O} = \{f \mid \int_M f^2 \, v_g = 1\}$ where v_g denotes the Riemannian volume element. The functional to be considered is

$$\phi(f) = \frac{1}{2} \int_M |df|^2_g \, v_g$$

that physicists would call the *energy of a free scalar field*. Since ϕ only involves first derivatives of f, it is natural to complete the space of geometric objects using the Sobolev norm whose square is the sum of the squares of the L^2-norms on the function and its differential. This space is usually denoted $H^1(M)$.

1.2.   The Euler-Lagrange equation of $\phi_{|\mathcal{O}}$ (directly derived from the First Variation Formula) is

$$\Delta^g f = d^{*g} df = \lambda^g f \ ,$$

where d^{*g} denotes the adjoint of d for the global scalar product, λ^g is called an *eigenvalue* of the Laplace-Beltrami operator Δ^g, the function f being called an *eigenfunction* associated to the eigenvalue. Therefore, f is a critical point and λ, which appears in the equation as a Lagrange multiplier, is a critical value of ϕ. (A general reference with interesting geometric examples is [19].)

1.3.   Without the L^2-constraint on functions, the smallest eigenvalue is 0 and the eigenfunctions are constant functions. The purpose of the constraint is therefore to get a non trivial eigenvalue and eigenfunction. This theory gives a generalization to all compact Riemannian manifolds of the theory of Fourier series on the circle S^1 with any Riemannian metric.

1.4.   Another way of approaching this problem is by considering a modified functional $\tilde{\phi}$ which is obtained by making it homogeneous of degree 0 w.r.t. the function by setting $\tilde{\phi}(f) = (\int_M |df|_g^2 \, v|g)/(\int_M f^2 \, v_g)$. This quotient is often called a *Rayleigh-Ritz quotient*.

If the manifold has a boundary, one has to define the boundary conditions. The most natural boundary conditions are the *Dirichlet* conditions, i.e., assuming that $f = 0$ on ∂M, or *Neumann* conditions, i.e., assuming that $\partial_n f = 0$ on ∂M where n denotes the unit normal field.

1.5.   *The Second Variation Formula of ϕ at a critical point f_0 is the quadratic form* Hess ϕ_{f_0} *defined on the space* $T_{f_0}\mathcal{O} = \{x \mid x \perp f_0\}$ *by*

$$(\text{Hess }\phi_{f_0})(x, x) = \int_M g^{-1}(df, df) \, v_g \,,$$

where g^{-1} denotes the metric on the cotangent bundle induced by g.

1.6.   The *index* at the critical point f_0 is the number of eigenvalues of Δ^g less than λ_0, and its *nullity* the *multiplicity* of λ_0, i.e., the dimension of the vector space $\{f \mid \Delta^g f = \lambda_0 f\}$.

B) A Refinement Due to J. Hersch

1.7.   A variant of this situation due to Joseph Hersch (cf. [70]) has nice geometric applications (cf. [71], [18], [31]).

1.8.   For $\mathcal{O}$, he takes the space of L^2-orthonormal bases of k-dimensional subspaces of the space $\mathcal{F}M$ of integral 0, or of its completion for the H^1-norm. The functional he is interested in is $\phi(f_1, \cdots, f_k) = \sum_{i=1}^k (1/\int_M g^{-1}(df_i, df_i) \, v_g)$. The supremum of ϕ is precisely $\sum_{i=1}^k (1/\lambda_i^g)$ where the λ_i denotes the ith eigenvalue of Δ^g (repeating eigenvalues as many times as necessary if they are multiple).

1.9.   One of the key features of ϕ is its conformal invariance when M has dimension 2. This means that the functional ϕ takes the same value for all Riemannian metrics $e^{2u} g$ where u is a function on M. This fact is clear since ϕ involves only Dirichlet integrals.

2. Harmonic Forms

In this section we deal with exterior differential forms. The basic operator we are interested in is again the Laplace-Beltrami operator $\Delta^g = d^{*g}\, d + d\, d^{*g}$. In A), we consider the spectral problem, and in B) we devote special attention to *harmonic forms*, i.e., forms lying in the kernel of Δ^g.

A) Eigenvalues on k-forms

2.1.   One takes $\mathcal{O}$ to be the unit ball for the L^2-norm in $\Omega^k M$, the space of exterior diferential k-forms, and

$$\phi(\omega) = \int_M (|d\omega|^2 + |d^{*g}\omega|^2)\, v_g \ .$$

Critical points of ϕ are again the *eigenforms* and critical values the *eigenvalues* of Δ^g.

B) Hodge Theory

2.2.   The space of geometric objects is isomorphic to $d\Omega^{k-1} M$, the image under the exterior differential of the space of exterior differential $(k-1)$-forms. One sees it as sitting as an affine space in Ω^k in the following way. One fixes $\alpha_0 \in \Omega^k M$ a *closed* differential k-form, and considers $\alpha_0 + d\Omega^{k-1} M \subset \Omega^k M$. Then, one sets $\phi(d\beta) = \int_M |\alpha_0 + d\beta|^2\, v_g$. A k-form β is critical if $\alpha = \alpha_0 + d\beta$ is *coclosed*, i.e., annihilated by d^{*g}, hence is *harmonic*.

2.3.   Moreover, the solution to this variational problem is always unique. (Indeed, the functional is by construction a non-negative quadratic functional defined on a vector space. Moreover, if $d^{*g}(\alpha_0 + d\omega_1) = d^{*g}(\alpha_0 + d\omega_2) = 0$, by setting $\omega = \omega_1 - \omega_2$, one gets $d^{*g}d\omega = 0$, hence $d\omega = 0$ after integrating by parts against ω proving that $d\omega_1 = d\omega_2$.)

2.4.   A Riemannian metric being picked, this procedure provides a natural representative for the cohomogical class defined by α_0. Indeed, de Rham theorem says that $H^k(M, \mathbf{R}) = \mathrm{Ker}d_k/\mathrm{Im}d_{k-1}$ where $H^k(M, \mathbf{R})$ denotes the cohomology space of degree k of the manifold M with real coefficients.

2.5.   An important feature of this construction is that it can be refined in a Kählerian setting (i.e., M being a complex manifold, and g a Hermitian metric whose associated skew 2-form is closed). Indeed, the algebra of exterior differential forms is naturally bi-graded by the complex type, and this bi-grading descends to harmonic forms. This provides a refined cohomological algebra, which can be put into correspondence with the cohomology with values in the sheaf of germs of holomorphic functions by the Dolbeault theorem.

3. Length and Energy of Curves

A) Geodesics as Shortest Paths

3.1.   We take $\mathcal{O} = \mathcal{C}_{p,q}M (= \{c \mid c : [\alpha, \beta] \longrightarrow M \ C^\infty, \ c(\alpha) = p, \ c(\beta) = q\})$, i.e., the space of all C^∞ curves parametrized on the interval $[\alpha, \beta]$ drawn on M and joining p to q.

A variant of this setting consists in working in a given homotopy class of closed curves drawn on M.

3.2.   Two functionals are classically considered on $\mathcal{C}_{p,q}M$, namely the *length* functional

$$\mathbf{L}(\gamma) = \int_\alpha^\beta \sqrt{g(\dot{c}(\tau), \dot{c}(\tau))} \, d\tau \ ,$$

and the *energy* functional

$$\mathbf{E}(c) = \frac{1}{2} \int_\alpha^\beta g(\dot{c}(\tau), \dot{c}(\tau)) \, d\tau \ .$$

Notice that the name *energy* is ambiguous here, since one can give to it two physical interpretations : $\mathbf{E}(c)$ can be viewed as the *kinetic energy* of a particle moving in the configuration space M in its movement along its trajectory c ; $\mathbf{E}(c)$ can also be viewed as the *elastic energy* contained in the string represented by the curve c.

3.3.   The First Variation Formula of $\mathbf{E}$ is just a reformulation of the celebrated Euler-Lagrange formula found in the XVIII th century and states that for a variation $s \mapsto c_s$ of the curve $c = c_0$

$$\frac{d\mathbf{E}(c_s)}{ds}_{|s=0} = - \int_\alpha^\beta g(D_{\dot{c}}\dot{c}, X) \, d\tau + [g(\dot{c}(\tau), X(\tau))]_\alpha^\beta \ ,$$

where $X = dc_s/ds_{|s=0}$ denotes a tangent vector at c to the space $\mathcal{C}_{p,q}M$ (which should be thought of as a vector field along the curve c), and D is the Levi-Civitá covariant derivative determined by the Riemannian metric g.

3.4.   Therefore, curves c which are critical points of $\mathbf{E}$, called *geodesics*, satisfy the equation $D_{\dot{c}}\dot{c} = 0$. In local coordinates (x^i), the equation is

$$\frac{d^2 c^i}{d\tau^2} + \sum_{j,k=1}^n \Gamma_j{}^i{}_k(c(\tau)) \frac{dc^j}{d\tau} \frac{dc^k}{d\tau} = 0 \ .$$

This differential equation is linear in the second derivative of the local expression of the curve, quadratic in the first derivative, and non-linear in the position of the curve through the local coefficients $\Gamma_{j}{}^{i}{}_{k}$ of the covariant derivative defined by $D_{\partial/\partial x^{j}}\partial/\partial x^{k} = \sum_{i=1}^{n} \Gamma_{j}{}^{i}{}_{k}\,\partial/\partial x^{i}$.

3.5.   Critical points of the length and energy functionals are intimately related : curves which extremize the energy functional are *geodesics* parametrized proportionally to arc length, whereas curves extremizing the length functional occur in families, the source reparametrization of an extremal curve being again an extremal curve. The minimizers of the energy functional are called (as they shoudl) *shortest paths* from p to q.

3.6.   A classical theorem states that *on a compact manifold, between any two points there is always a minimizing geodesic.*

3.7.   The Second Variation contains a lot of geometric information on the metric g. If c is a geodesic, and $X \in T_{c}\mathcal{C}_{p,q}M$,

$$\mathbf{Hess}\mathbf{E}_{c}(X, X) = \int_{\alpha}^{\beta} g(D_{\dot{c}}D_{\dot{c}}X + R_{X,\dot{c}}\dot{c}, X)\,d\tau \ .$$

Hence, the *nullity* of $\mathbf{E}$ is made of the vector fields X along a geodesic c satisfying the *Jacobi equation* $D_{\dot{c}}D_{\dot{c}}X + R_{X,\dot{c}}\dot{c} = 0$, called *Jacobi fields*. If there exists some non everywhere vanishing Jacobi field $X \in T_{c}\mathcal{C}_{p,q}M$, one says that the points p and q are *conjugate along* c. The *index* of a geodesic c between two points p and q is zero when c is minimizing. It can only be non-zero when between p and q one can find a point conjugate to one of the two end points along c.

These ingredients are basic in the extension to the infinite dimensionsal manifold $\mathcal{C}_{p,q}M$ of Morse theory, the function into consideration being the energy functional.

A typical application of this is the existence of infinitely many closed geodesics on any compact Riemannian manifold. The difficulty lies in the fact that all curves obtained in traveling a closed geodesic more than one time are again geodesics, and should be discarded as providing non geometrically distinct curves. The final solution of the existence of infinitely many closed geodesics on the 2-sphere for any Riemannian metric has been proved by V. Bangert and R. Frank.

3.8.   Conjugate points are used in numerous ways in Riemannian geometry. We just quote here a few basic theorems whose proof relies on this concept.

SYNGE'S THEOREM.   *Any orientable even-dimensional compact Riemannian manifold with positive sectional curvature is simply connected.*

HADAMARD-CARTAN'S THEOREM. *Let (M, g) be a complete Riemannian metric with negative sectional curvature and $p \in M$. The exponential map $\exp_p$: $T_p M \longrightarrow M$ which to a tangent vector v at p associates $\exp_p(v)$, the point at time 1 on the geodesic issued from p with velocity vector v at time 0 is a covering map.*

B) An Energy for Knots

3.9.   If one takes $(M, g) = (\mathbf{R}^n, e)$ where e stands for a Euclidean metric, then geodesics are straight lines. But one can use the energy functional to define an interesting functional on *knots*, i.e., maps c from $\mathbf{S}^1$ to $\mathbf{R}^n$. For $\alpha, \beta \in \mathbf{S}^1$ and a knot c, one sets

$$d_c(\alpha, \beta) = \min \left\{ \int_\alpha^\beta |\dot{c}(\tau)| \, d\tau , \int_\beta^\alpha |\dot{c}(\tau)| \, d\tau \right\} .$$

One can then introduce the functional $\mathcal{F}$ on $\mathcal{C}(\mathbf{S}^1, \mathbf{R}^n)$ by setting

$$\mathcal{F}(c) = \int \int_{\mathbf{S}^1 \times \mathbf{S}^1} \left(\frac{1}{|c(\alpha) - c(\beta)|^2} - \frac{1}{d_c(\alpha, \beta)^2} \right) |\dot{c}(\alpha)| \, |\dot{c}(\beta)| \, d\alpha \, d\beta .$$

The functional $\mathcal{F}$ is independent of the parametrization (i.e., it depends only on the geometric shape of the knot), and conformally invariant (i.e., this makes it independent of the size of the knot) (cf. [65]). It is suggested by a biological model, where the knot is a complicated molecule which may be knotted. (For another approach, cf. [85].)

3.10.   Some toric knots in $\mathbf{S}^3$ (sitting as a round sphere in $\mathbf{R}^4$, a geometric setting which is consistent with the conformal invariance of $\mathcal{F}$) are critical points of $\mathcal{F}$. They can be put into correspondance with orbits of some group actions (cf. [81]).

4. The Energy of Maps

Although historically the energy of curves has been studied for much longer, the notion of energy of maps has been studied very extensively in the last twenty years (for reports, see [52], [53], [54]). There are many reasons for this :
   – basic existence and regularity theorems have become available ;
   – new phenomena, connected to models of interest to physicists, appear when the source space is 2-dimensional because of an extra conformal invariance in this dimension ; these phenomena, often called *bubbling off* phenomena, have been completely clarified through a powerful geometric analysis ;

– a great variety of geometric situations have been found in which the selection of a special map could be done by taking it to be an extremal of the energy functional or a modification of it.

A) Harmonic Maps

4.1.   Let (M, g) and (N, h) be two Riemannian manifolds of respective dimensions m and n. We take $\mathcal{O} = \{f \mid f : M \longrightarrow N \ C^\infty\}$. One sets

$$\mathbf{E}_{g,h}(f) = \frac{1}{2} \int_M g^{-1} . f^* h \, v_g$$

$$= \frac{1}{2} \int_M \left(\sum_{i,j=1,\alpha,\beta=1}^{m,n} g^{ij} \frac{\partial f^\alpha}{\partial x^i} \frac{\partial f^\beta}{\partial x^j} h_{\alpha\beta} \right) v_g \ , \ cr$$

where g^{-1} denotes the metric on the dual space induced by g whose local expression is the inverse matrix to that of g, and . the contraction for the intrinsic expression, and where we have used local coordinates (x^i) on the source manifold M and local coordinates (y^α) on the target manifold N.

4.2.   Critical points of the energy functional $E_{g,h}$ are called *harmonic maps*. The local condition that a map f must satisfy in order to be harmonic is

$$(\Delta f)^\alpha = \sum_{i,j=1}^{m} g^{ij} \left(\frac{\partial^2 f^\alpha}{\partial x^i \, \partial x^j} - (\Gamma^{(g)})_i{}^k{}_j \frac{\partial f^\alpha}{\partial x^k} - (\Gamma^{(h)})_\beta{}^\alpha{}_\gamma \frac{\partial f^\beta}{\partial x^i} \frac{\partial f^\gamma}{\partial x^j} \right) = 0 \ .$$

One notices that, unless the source metric is flat (i.e., the Christoffel symbols $\Gamma^{(g)}$ of the metric g can be taken constant by an appropriate choice of coordinates), the equation is quadratic in the first derivative of the map f. The non-linearity of the system of partial differential equations, which is quasi-linear elliptic because of the leading term, comes also from the fact that, one more time unless the metric on the target space N is flat, the Christoffel symbols of h are taken at the point $f(p)$.

4.3.   Another way of expressing harmonicity of a map is to say that df is harmonic, when viewed as a $f^* TN$-valued 1-form for the pulled back connection deduced from the Levi-Cività covariant derivative of the metric h.

4.4.   We skip the Second Variation Formula which can be found in [92] and [128]. The main new feature concerning it is the simultaneous presence in the formula of the Ricci curvature of the source manifold M and of the full curvature tensor (or the sectional curvature) of the target manifold N. There is nothing to be surprised about since the differential of f is a 1-form on M, hence an object sensitive to the Ricci curvature of the source space M. (For the use of stable harmonic maps, cf. [99].)

4.5.   The first important existence theorem is due to Eells and Sampson (cf. [55]) when the target space has non-positive curvature. (For other theorems using flow methods, cf. [129] and [37].) Many other theorems have been found later. Notice though that the existence theory is very subtle. Eells and Wood showed for example that there is no harmonic map of degree 1 from $\mathbf{T}^2$ to $\mathbf{S}^2$.

4.6.   The basis of the regularity theory is due to Schoen-Uhlenbeck (cf. [118]). One already encounters some problem, when one is to define the space of H^1 maps from a manifold to another one. Two definitions seem natural. One consists in taking the completion of the space of C^∞ maps for the H^1-norm. The other one, which is very tempting because one is not forced a priori to work in charts, is to embed isometrically N into a Euclidean space (this is possible by the Nash theorem), so that one can work in the vector space of H^1-maps from M to $\mathbf{R}^n$. One then retains only maps which almost everywhere take their values in the submanifold N. The main difficulty is that these two definitions do not coincide. This question has been settled recently (cf. [24]). As usual, minimizers are more regular than other critical points.

For the most recent results on the regularity theory of harmonic maps (and the study of their singularities), consult the notes by Leon Simon in this volume.

4.7.   When M is 2-dimensional, the energy functional is very analogous to a Dirichlet integral, hence does depend only on the conformal class of g. This means that when M is orientable, the functional E is attached to its structure as a Riemann surface.

This conformal invariance (which can also be interpreted as a scale invariance) is responsible for the lack of compactness that one encounters when M is 2-dimensional. From an analytic point of view this is reflected in the fact that H^1-maps are not necessarily continuous, but fall in the limiting case of Sobolev embedding theorems. One of the consequences is that a sequence of maps may converge weakly to a map belonging to another homotopy class (cf. [114]). This lack of compactness can be pictured as a bubbling off phenomena (to suggest that a sphere may concentrate at a point).

4.8.   An interconnection with the theory of complex manifolds has been used in higher dimensions. The basis of these generalisations is that holomorphic or anti-holomorphic maps are harmonic. In order to prove that two complex manifolds are biholomorphically equivalent, one has to show that a harmonic map between them is holomorphic, and finally that it realizes this equivalence.

B) Prescribed Singularities

4.9.   In recent years, special attention has been devoted to maps u from the 3-ball B^3 to the 2-sphere S^2 because of the possible interpretation of such maps as models of liquid crystals. Since the source space has a boundary, one of course needs to fix some boundary conditions, in this case a map $\varphi : S^2 \longrightarrow S^2$. The homotopy class of φ is determined by its degree. We shall assume in later developments that φ has degree 0.

4.10.   The equation that a weakly harmonic map u satisfies is

$$\Delta u + u \, |du|^2 = 0 \ .$$

Such maps may have singular points (the source space is 3-dimensional). To each singular point, one can attach an *index* which is the degree on neighbouring spheres. For a map f, Béthuel, Coron et Lieb use a *relaxed energy* (cf. [25], [69]) which can be defined as follows

$$\mathbf{E}_\lambda(f) = \mathbf{E}(f) + 8\pi \, \lambda \mathbf{L}_0(f)$$

with

$$\mathbf{L}_0(f) = \inf\{\mathbf{L}(c) \mid \partial c = \sum_{b \in \mathrm{Sing}u} (\deg u)(b)\, \delta_b\} \ .$$

5. Minimal Submanifolds

The study of surfaces in Euclidean 3-space has been the root of most of the concepts of Differential Geometry. The theory has developed much beyond this original motivation.

A) The Area Functional

5.1.   Let M be an m-dimensional manifold and (N, h) a Riemannian n-dimensional manifold.

We take $\mathcal{O} = \{f \mid f : M \longrightarrow N \text{ of class } C^\infty\}$. We consider the *area functional* ϕ

$$\phi(f) = \int_M v_{f^*h} \ .$$

5.2.   It is a very classical result that critical points of ϕ are *minimal immersions*, i.e., immersions whose mean curvature vector vanishes, in other words for which the second fundamental form B, a normal bundle valued symmetric bilinear form, is tracefree. In the case of surfaces in $(\mathbf{R}^3, e)$ (and more generally for hypersurfaces), this means that the immersed surface has opposite curvature radii, i.e., has a saddle shape.

When $m = 2$, the theory of minimal surfaces is intimately related to that of harmonic maps because the two functionals are in this case quadratic in the map. In fact, a harmonic map which is also critical with respect to the variations of the conformal structure is minimal, cf. [114]. It is also of practical interest because of its relation to soap films as noticed by the Belgian physicist *Plateau* in the middle of last century (cf. [109]), hence the name of *Plateau problem* given to the search of a minimal surface spanning a wire having a definite shape in the ambient space. The mathematical solution of this problem was not found until the late 30s (cf. [49] and [113]). Even for simple wires, singularities may occur, cf. [132].

In this dimension the relation of this problem to complex function theory has been recognized more than a century ago. The capital fact is the famous Enneper-Weierstrass representation of minimal surfaces as the real part of three indefinite integrals involving complex functions which can be interpreted geometrically in terms of the Gauß map of the surface.

5.3.   The Second Variation Formula is also classical. We state it for a normal vector field V along $f(M)$, a normalized infinitesimal variation, using the normal Laplace operator $\Delta^N = D^{N*}D^N$ and the induced metric $g = f^*h$,

$$
\mathrm{Hess}_f \phi(V, V) = \int_M \left(g(\Delta^N V, V) - \sum_{i,j=1}^m (B_{e_i, e_j}, V)^2 + \sum_{j=1}^m g(R^N_{e_j, V} e_j, V) \right) v_g.
$$

One should point out that the differential operator acting on infinitesimal variations is again a Fredholm operator on the space of normal vector fields along $f(M)$.

5.4.   In the late 70's, minimal submanifolds have been used to explore geometric situations, in particular when $m = 2$. Stable minimal surfaces have been used for example in the proof by R. Schoen and S.T. Yau of the positive mass conjecture (cf. [119], [121]). (For applications to the topology of 3-manifolds, see [93] and [120].)

5.5.   The theory has also a very developed measure-theoretic side, which grew into a theory of its own, the so-called *Geometric Measure Theory*. A comprehensive textbook for that side of the theory is [60]. (For approaches inspired by this point of view with a direct geometric bearing, one can also consult [2].)

This side is necessary to deal with singularities which may occur even with stable minimal submanifolds when the dimension is at least 7. (For the first ground-breaking article, cf. [127], [26], and for a survey of this period, [86] and of later periods [123].)

5.6.   Minimal surfaces have also been used with great success in the late 70s as geometric tools to investigate the topology of manifolds satisfying local curvature assumptions (cf. [120]). This approach enabled S.T. Yau and R. Schoen to solve the *positive mass conjecture*, one of the challenging problems in General Relativity of isolated universes (cf. [119], [121]).

5.7.   In recent years, many new examples of embedded minimal surfaces of finite total curvature (cf. [79]) have been found, sometimes with the help of computer images to suggest the right controls avoiding self-intersections at finite distance. This wealth of new examples provided interesting models for various sciences, from theoretical physics to applied chemistry (cf. [3]) or biology (cf. [7]). (For the study of free boundary interfaces, cf. [38].)

B) Surfaces of Constant Mean Curvature

5.8.   It is a classical (and easy) result that there exists no compact minimal surface without boundary in $\mathbf{R}^3$. It is therefore natural to look for a modification of the variational problem of the area functional in order to include such surfaces among the extremal points.

One achieves that by putting the constraint that the volume enclosed by the immersed surfaces be fixed. This modifies the space of geometric objects on which one considers the functional. The constraint given by the total volume introduces a Lagrange multiplier in the Euler-Lagrange equation, and the critical points are *surfaces with constant mean curvature*. (The Lagrange multiplier is precisely the constant value of the mean curvature .)

Constant mean curvature surfaces model soap bubbles.

5.9.   The theory of constant mean curvature surfaces has only been recently attacked systematically. Among the difficult conjectures concerning them, one should quote Hopf asking whether the round sphere is the only compact constant mean curvature surface in $\mathbf{R}^3$. This was disproved by H. Wente (cf. [141]) who established that the 2-torus can be immersed with constant curvature. His purely PDE approach has been later greatly amplified (and simplified) by U. Abresch and U. Pinkall when properly put in relation with completely intergrable systems.

C) Bending and Stretching of Surfaces

5.10.   One can also focus one's attention on other functionals than the area. In particular when one is interested in constructing realistic models of membranes, one considers both their *bending* and their *stretching*. Membranes are known to play an important in many Natural Sciences.

This leads to the study of families of functionals, indexed by parameters that one tunes for a specific problem, such as

$$\phi_{a,b,c}(f) = \int_M \left(a + b\, H_f^2 + c\, G_f \right) v_{f^\bullet e} \,,$$

where H_f denotes again the mean curvature of the surface and G_f its Gaussian curvature (the determinant of the second fundamental form which, by Gauß' theorema egregium, is known to be accessible from the intrinsic Riemannian geomery of the surface as defined by its first fundamental form $f^* e$).

5.11. If one does not put any constraint (such as a fixed total volume), the Euler-Lagrange equation is

$$\Delta H_f = 2\, H_f^2 - \left(G_f - \frac{a}{b} \right) H_f \,.$$

The absence of c in the formula reflects the Gauß formula, stating that $\int_M G_f\, v_{f^\bullet e}$ does not not depend on f.

It is also interesting to note that the functional $\int_M H_f^2\, v_{f^\bullet e}$ is conformally invariant. Getting the optimal lower bound for this functional on the torus is known as the *Willmore conjecture*, cf. [145], [89] and [32].

6. Yang-Mills fields

A) The General Theory

6.1. The framework of Yang-Mills theory has been introduced by physicists in an effort to give classical models for quantum theories of elementary particles. One fixes a compact Lie group G (thought of as the symmetry group of the theory). There are several equivalent formalisms to present the theory. Here, we use the vector-bundle formalism. Ones takes $\mathcal{O} = \mathcal{A}E$, the space of G-connections of a vector bundle $E \longrightarrow M$. One assumes that the base space M is endowed with a Riemannian metric. The space $\mathcal{A}E$ is an affine space modelled on the vector space of differential 1-forms taking their values in the bundle of infinitesimal G-automorphisms of E.

The Yang-Mills functional is

$$\phi(\nabla) = \int_M |F^\nabla|^2\, v_g \,,$$

where F^∇ is the *curvature* of the connection ∇ (physicists speak of the *field strength*.) It is defined, for $\nabla = \nabla_0 + A$ with ∇_0 a reference connection, as

$$F^\nabla = F^{\nabla_0} + d^{\nabla_0} A + [A \wedge A] \,.$$

Here, d^∇ denotes the exterior differential for vector-valued differential forms using the covariant derivative ∇. Note that the norm on the curvature uses both the metric on the manifold M and a biinvariant metric on the Lie algebra of G.

6.2.   The Euler-Lagrange equation for the Yang-Mills functional is

$$d^{\nabla *} F^\nabla = 0 \ ,$$

where $d^{\nabla *}$ denotes the adjoint of d^∇ for the L^2-scalar product on vector-valued 2-forms.

This system of partial differential equations is non-linear as soon as the group G is non-Abelian, as shown in the formula defining the curvature. Critical points of the Yang-Mills equations are called *Yang-Mills connections*.

6.3.   Since F^∇ is automatically closed (this is the so-called *second Bianchi identity*), the curvature of a solution of the Yang-Mills equations is a harmonic 2-form, hence the fact that, from a purely mathematical point of view, Yang-Mills theory is often quoted as a non-linear Hodge Theory.

6.4.   The Second Variation of ϕ is given, for an infinitesimal deformation A of the connection ∇, by

$$\mathrm{Hess}_\nabla (A, A) = \int_M (d^{\nabla *} d^\nabla A + \sum_{i=1}^n [R^\nabla_{.,e_i}, A_{e_i}], A) \, v_g \ .$$

We note that the second variation is not a priori an elliptic operator, hence presents some degeneracies. There is a deep reason for that. Indeed, the Yang-Mills functional is invariant under the action of the *gauge group*, i.e., the group $\mathcal{G}E$ of sections of the automorphism group of E, which is an infinite dimensional group. As a result, when a connection ∇ is critical, all connections in the orbit $\mathcal{G}E.\nabla$ are critical. One can turn the Second Variation into a Fredholm operator by restricting the infinitesimal variation to being transversal to the orbit, i.e., by imposing the condition $d^{\nabla *} A = 0$. (For a general discussion of the variational side, cf. for example [27].)

B) The Four-Dimensional Case

6.5.   Both for physical and mathematical reasons, the Yang-Mills functional has very special features in dimension 4, hence this special section. The theory is *conformally invariant*, which means that the functional depends only on the conformal class of the metric on M. This is in many ways analogous to the 2-dimensional situation for harmonic maps, the skew-field of quaternions playing here in some sense the role played by the field of complex numbers there.

6.6.   The crucial special features in 4 dimensions are the existence of a topological lower bound for the functional, and the existence of special minimizing solutions.

The formulation of both requires the use of the *Hodge map* $*$, which, on an n-dimensional oriented manifold M, to a k-form associates an $(n-k)$-form. Precisely, for $\alpha \in \Omega^k M$, $\beta \in \Omega^{n-k} M$, $*\alpha$ is defined by $(*\alpha, \beta)\, v_g = \alpha \wedge \beta$.

For $n = 4$, $*$ is a self-map of the space $\Omega^2 M$ of 2-forms, in fact an involution with invariant subspaces $\Omega^+ M$ and $\Omega^- M$ corresponding to the eigenvalues 1 and -1. Elements of $\Omega^+ M$ are called *self-dual* forms, resp. of $\Omega^- M$ *anti-self-dual* forms.

In this context, one has

$$4\pi^2 |p_1(E)| \le \phi(\nabla) \ ,$$

where $p_1(E)$ denotes the first *Pontryagin number* of the bundle E. Moreover, *equality occurs if and only if F^∇ is self-dual or anti-self-dual.*

6.7.   From this inequality, one sees that self-dual connections are absolute minima of the Yang-Mills functional, hence are critical points. The construction of self-dual connections for special manifolds M, such as the 4-sphere or the complex projective plane, can be translated into a complex analytic problem concerning a complex manifold associated to M, its *twistor space*. This approach, initiated in [33], has linked in an interesting way this variational theory and analytic geometry. (A good general reference for this point of view is [9].)

6.8.   After it was proved that stable critical points are absolute minima when the group G is not too large (e.g., SU_2 or SU_3) (cf. [29], [30]), the search for non-minimal critical points on manifolds as simple as the 4-sphere has finally been successful, cf. [108], [115] and [124].

6.9.   When (M, g) is a general Riemannian manifold, one cannot anymore construct explicit solutions, and one has to rely on techniques from the theory of Partial Differential Equations. The main tool to prove existence of self-dual solutions is a grafting procedure introduced by C. Taubes, cf. [130]. This process can be applied when the intersection form of M is positive definite. It exploits the conformal invariance of ϕ, and the explicit solutions on the 4-sphere.

Solutions occur in moduli and form a space whose structure, when properly analyzed, provides very useful information on the differential topology of M already for the simplest non-Abelian Lie group SU_2. Crucial instruments for that are compactness theorems due to K. Uhlenbeck describing how a sequence of self-dual connections can escape to infinity in the moduli space (cf.

[134], [135], [136], [98] ; for general presentations, cf. [87] and [64]). This is the starting point of the new approach to 4-dimensional Differential Topology introduced by S. Donaldson (cf. [40], [41], [42], [43], [44], [45], [46], [47], and the comprehensive book [48]).

C) The Chern-Simons Functional

6.10.   When M is 3-dimensional, another theory (also suggested by physicists) has been the source of interesting topological developments. The functional is called the *Chern-Simons* functional and is defined as follows. One takes $\mathcal{O} = \mathcal{A}E$ as space of geometric objects, and

$$\phi(\nabla) = \frac{1}{8\pi^2} \int_M \text{Trace}\left([dA \wedge A] + \frac{2}{3}[[A \wedge A], A]\right) .$$

The functional comes from the transgression of the curvature expression of the first Pontryagin form of M, which vanishes for dimensional reasons.

6.11.   The First Variation formula is

$$d\phi(\nabla) = -\frac{1}{4\pi^2} F^\nabla$$

so that the critical points are flat connections.

6.12.   The Second Variation of ϕ is given by the differential operator $*d^\nabla$, which, being non elliptic, has a priori no well defined index. The necessary modification to get useful information also in this context has been proposed by A. Floer, cf. [63]. (For more on this theory, one can consult the notes of the series of lectures by K. Fukaya.)

7. The Total Scalar Curvature Functional

The curvature field R_g of a metric g is a 4-tensor field, often called the *Riemann-Christoffel tensor*, which is quite complicated. It is therefore natural to look for simpler invariants. Its simplest form is the *scalar curvature*, obtained by contracting R_g against the metric in the appropriate places, and that we denote by Scal_g. (Because of its many skew-symmetries, there is up to sign only one way of contracting it fully.) The total scalar curvature functional is a priori defined on the space $\mathcal{O} = \mathcal{M}M$ (where we assume M to be compact) by

$$\phi(g) = \int_M \text{Scal}_g \, v_g .$$

In fact there are various situations where this functional has been studied quite systematically.

A) The Yamabe Problem

7.1.   We begin by the more drastic reduction, consisting in looking only at metrics belonging to a given *conformal class*. The space of conformal metrics is $\mathcal{O} = \{g \mid c^2 u\, g_0\}$ for a fixed metric g_0. When the dimension $n \geq 3$, it turns out to be more convenient to write the conformal factor $e^u = f^{4/(n-2)}$ with the extra condition $f > 0$. The formula for the scalar curvature is

$$f^{\frac{n+2}{n-2}} \operatorname{Scal}_g = 4\,\frac{n-1}{n-2}\Delta_{g_0} f + \operatorname{Scal}_{g_0} f \ .$$

The *normalized* total scalar curvature is

$$\phi(f) = \frac{4\,\frac{n-1}{n-2} \int_M |df|^2\, v_{g_0} + \int_M \operatorname{Scal}_{g_0} f^2\, v_{g_0}}{\int_M f^{\frac{2n}{n-2}}\, v_{g_0}} \ .$$

7.2.   The critical points of ϕ are *metrics with constant scalar curvature*, cf. [151]. The subtlety of the theory lies in the fact that the functional involves the $L^{2n/(n-2)}$-norm, hence falls in the limiting case of Sobolev inequalities. The main consequence for the variational theory is the possibility for a minimizing sequence to converge to 0. One therefore needs to refine the estimate, in fact one has to go as far as proving an optimal Sobolev inequality. Most cases have been handled by T. Aubin in [14] (see also [133]), and the final most subtle ones by R. Schoen, cf. [116], and [117]. (For a survey, cf. [88].)

7.3.   There are geomeric reasons for the lack of compactness of the functional. Indeed on the standard sphere the constant curvature metrics form a non-compact set in a conformal class because of the action of the non-compact group of conformal transformations. The whole point of the theory is to show that this phenomenon is typical of the sphere and of concentration phenomena that the group of conformal transformations of the sphere can create.

B) Einstein Metrics

7.4.   We now come to the case of general metrics. Interest for the total scalar curvature came from its use to provide the field equations of General Relativity by Einstein and Hilbert (cf. [58] for the general setting without the field equations and [57] and [72] for the complete theory). Of course, in this context the metric is Lorentzian and not Riemannian, but the calculation leading to it takes the same form.

7.5.   Because of the sensitivity to dilations of the total scalar curvature functional, it is natural to introduce a normalizing constraint in the variation problem by introducing the constraint $\Gamma(g) = \int_M v_g - 1$. The space of geometric objects that we chose therefore consists of metrics with total volume 1.

The First Variational Formula reads

$$d\phi(g) = -\mathrm{Ric}_g - \frac{1}{2}\mathrm{Scal}_g\, g\ ,$$

where Rc_g denotes the *Ricci curvature* a symmetric 2-tensor field which is, up to sign, the only non trivial tensor that one obtains by contracting once the Riemann-Christoffel tensor.

When one takes into account the Lagrange multiplier coming from the constraint Γ, one gets for critical points the solution of the system of Einstein equations, the so-called *Einstein metrics* defined by

$$\mathrm{Ric}_g = \frac{1}{n}\, g\ .$$

(For a systematic presentation of these metrics, cf. [21].) From a PDE point of view, there is of course a main difference between these two systems when the metric is Lorentzian or Riemannian. When conveniently reduced, i.e., in appropriate system of coordinates (the so-called *harmonic* coordinates), the system of equations is hyperbolic in the first case whereas the system of equations is elliptic in the Riemannian setting.

7.6.   It is a consequence of the second Bianchi identity on the curvature that the scalar curvature of an Einstein metric is automatically constant, the so-called *Einstein constant*.

By scaling the metric, it is possible to reduce the Einstein constant to three possible values : 1, 0 and -1.

7.7.   The normalization of the coordinates which is referred to above points to the fact that the system of Einstein equations is intrinsic. (Physicists would say *covariant*.)

This invariance explains why the system of Einstein equations is degenerate, and also why the Second Variation has to be considered only for infinitesimal variations transversal to the action of the group of diffeomorphisms on the space of metrics. This can be achieved by considering symmetric differential 2-forms h which are divergence-free. For such infinitesimal variations h, the Second Variational Formula at an Einstein metric g with constant λ reads

$$\mathrm{Hess}_g(h,h) = \int_M (D^*Dh - 2\lambda\, h + R(h), h)\, v_g\ ,$$

where D denotes the Levi-Cività connection of the metric g, and R denotes the action of the curvature on symmetric 2-tensors.

7.8.   The study of moduli spaces of Einstein metrics has greatly advanced in the late 80's (for earlier work, cf. [82]), thanks to theorems establishing possible limits of sequences of metrics with controlled Ricci curvature (such as Einstein metrics). (One can consult for example [97], [67], [4], [5], [6], [15], and [16].)

B) Some Other Functionals of the Curvature

7.9.   The study of other functionals on the space of metrics connected to curvature was also stimulated by General Relativity. Among them are functionals involving the curvature quadratically. By a theorem of Hermann Weyl, they depend only on three real parameters

$$\phi_{a,b,c}(g) = \int_M (a\,|R|^2 + b\,|\mathrm{Ric}_g|^2 + c\,\mathrm{Scal}_g^2)|_g \ .$$

It is interesting to note that C. Lanczos (cf. [84]) discovered in 1938 that this functional $\phi_{1,-4,1}$ has a differential which is identically zero in dimension 4. In fact this establishes that this integral is a differential invariant of the manifold M. It is even a topological invariant, namely $8\,\pi^2\,\chi(M)$, where $\chi(M)$ the Euler-Poincaré charasteristic of M.

7.10.   First Variational Formulas for such functionals are quite complicated, and it is difficult to discuss their critical points in great generality. This is even worse for Second Variational Formulas of course.

7.11.   Other geometries which are more special than Riemannian geometry also use variational problems to select special metrics. This is in particular the case for Kählerian Geometry.
    Although, when the Kähler classe is fixed, the total scalar curvature of a Kähler metric does not depend on the metric in its class (cf. [20] for a systematic account), the functional $\phi_{0,0,1}$ has been studied extensively by E. Calabi (cf. [34] and [35]). In fact one can prove that for Kählerian surfaces (i.e., for manifolds of real dimension 4), all quadratic functionals of the curvature have the same variational theory because they are tied by two linear relations, one connected with the Euler characteristic, and a Kählerian invariant.

7.12.   Another functional of the curvature which seems natural is

$$\phi(g) = \int_M |R_g|^{\frac{n}{2}}\,v_g$$

because of its invariance under dilations.

   All preceding examples of variational theories involve integrals of locally defined quantities.

8. Some Other Non-Local Functionals of Riemannian Metrics

    In this section, we consider a few functionals for Riemannian metrics which are not defined by integrals of local expressions in the metric. They are global in some sense, although some of them fail to be local only in a subtle way. We will work with $\mathcal{O} = \mathcal{M}_\infty M$, the space of Riemannian metrics of the manifold M with total volume 1 (to avoid worrying about normalizations).

A) Eigenvalues as Functions of the Metric

8.1.   We introduced the eigenvalues of the Laplace-Berltrami operator Δ_g earlier in this chapter, and we consider them as functions of the metrics. Other operators are also of interest in this respect, such as the family of Schrödinger operators $\Delta_g + V$ where V is a potential. Among those, one plays a specific role, namely the *Yamabe operator*, $L_g = 4(n-1)/(n-2)\Delta_g + \text{Scal}_g$ because of its conformal covariance. (This is why it is also called the *conformal Laplacian*.)

8.2.   One of the difficulties known to appear with functionals such as $\phi(g) = \lambda_g$ is the lack of differentiability of ϕ at metrics where the eigenvalue under scrutiny has a variable multiplicity. This is the cause of a bifurcation.

Therefore, the First Variational Formula can be established only at points where such problems do not occur. This is ensured for example if the eigenvalue is simple. If f_g is the eigenfunction associated to the eigenvalue λ_g, one then has

$$d\phi(g)(h) = \int_M \frac{d}{dt}\Delta_{g+th|t=0} f_g\, f_g\, v_g \;,$$

hence involves the variation of the operator as it acts on the eigenfunction associated with λ_g. This is a general principle.

Second Variational Formulas for the eigenvalues of the Laplacian can be found in [18], and variation formulas for the eigenvalues of the Dirac operator in [28].

8.3.   More complicated functionals of the eigenvalue can also be considered. M. Berger considers for example for any fixed integer k, the functional $\phi_k(g) = \text{Vol}_g^{-2/n} \sum_{i=1}^k \lambda_i^{-1}$ where the eigenvalue λ_i denotes the i-th eigenvalue of the Laplacian.

Critical points of ϕ_k correspond to isometric embeddings of M by evaluation in the k-dimensional space generated by the first k eigenfunctions. (This is discussed in [18].)

8.4.   A nice geometric application of this formula for the conformal Laplacian is the following theorem.

THEOREM (Kazdan-Warner). *Any metric with vanishing scalar curvature on a manifold admitting no metric with positive scalar curvature are Ricci-flat.*

B) Regularized Determinants

8.5.   There are even more complicated functionals of eigenvalues known as *regularized determinants*, denoted det'. They have been considered systematically by physicists in the *renormalization theory*. On a manifold M, one defines $\det'\Delta_g = e^{-\zeta'(0)}$ where the ζ-function is defined as $\zeta(s) = \sum_{i=1}^{\infty} \lambda_i^{-s}$, and is proved to be holomorphic for $\Re s$ sufficiently large.

The First Variational Formula for det' has been derived by S. Polyakov, and can be found in [100], [101] and [102]. On a Riemann surface, critical points are constant curvature metrics. The theory is even more interesting for surfaces with boundary. The higher dimensional case is much more involved.

8.6.   Another type of functionals is closely connected to the previous one : the so-called η-invariants. They are associated to first order differential operators, such as the operator $d + d^*$ on exterior differential forms which are square roots of Laplacians. The eigenvalues of such operators are of both sign. η-invariants measure the balance between positive and negative eigenvalues. They are very global in their nature, but their variations are local.

C) Metric Entropy

8.7.   A completely different type of functionals involving more geometric quantities can be associated to metrics. One quantity which has captured a lot of attention in recent years is the *entropy*, in particular in connection with the study of manifolds with negative curvature. This notion is directly connected to the asymptotic behaviour of the geodesic flow. (For details, the reader can consult [22] and [23].)

8.8.   Very important geometric results have been deduced from the characterization of symmetric metrics on compact quotients of symmetric spaces as minima of the entropy, cf. [23].

Chapter III

Symmetry Considerations, Topological Constraints, and Interactions with Physics

1. Symmetry Considerations

We begin this section by considering in Sub-Section A the simplest case of symmetries, i.e., when a group acts on the space of geometric objects and preserves the functional. When the group of invariance is infinite dimensional, this has deep consequences on the Euler-Lagrange equations of the Geometric Variational Problem. In Sub-Section B, we recall the Principle of Symmetric Criticality, and explain some of its consequences. We devote the next Sub-Section C to Conservation Laws that represent a subtle way in which symmetry considerations can affect a Geometric Variational Problem, in particular if the group involved is non compact.

A) Group Actions

1.1.   It is a classical technique to study spaces where the description of objects can be reduced to algebraic considerations. This is often the case on *homogeneous* spaces, i.e., spaces on which a group is acting transitively. This most of the time allows to reduce the search for critical elements to a finite dimension test space since invariant objects are determined by their value at a point, hence belong to a finite dimensional vector space. (For general considerations, cf. [78].)

Since the most common spaces are indeed homogeneous, many of the canonical examples of geometric situations are indeed of this type. (For left-invariant Einstein metrics on Lie groups, cf. [77], for invariant Yang-Mills equations, cf. [138] and [74].)

1.2.   When the group does not act transitively, but has orbits of codimension k, it is often possible to push down the variational problem under consideration onto the space of orbits (a k-dimensional space in general with singularities) and get another variational problem with less dimensions. (For an application to the search for minimal submanifolds, cf. [73].)

B) The Principle of Symmetric Criticalily

1.3.   When an object which competes for being a critical point of a variational problem is invariant under a group acting on the situation, it is tempting to test it only against invariant infinitesimal variations.   This is the so-called Symmetric Criticality Principel, cf. [106]. This principle is not valid only if the group is compact. The main advantage of course is to reduce the problem to a finite dimensional extremal problem.

2. Topological Constraints

A) Homology and Cohomology

2.1.   In the variational problems that we considered, we met a number of times topological constraints.

    This is of course the case when the space of geometric objects is defined using a topological constraint, i.e., curves of a given homotopy class, applications having a certain degree, closed differential forms belonging to a given cohomology class, Kähler forms belonging to a given Kähler class. This is connected to the fact (that we already mentioned) that a variational problem can be of great help in selecting a geometrically interesting object in topologically distinguished set.

2.2.   We leave the appearance of topological constraints in the functional itself for the next sub-section.

B) Characteristic Classes and Curvature Integrands

2.3.   One of the most striking result in Global Differential Geometry is the possibility of expressing global topological invairants by local means. The very basic example is of course the Gauß-Bonnet formula expressing the Euler characteristic $\chi(M)$ of a surface M

$$2\pi \chi(M) = \int_M \sigma_g\, v_g$$

where σ_g denotes the *Gaussian curvature* of any Riemannian metric g on M.

    This has been widely generalized through the Chern-Weil theory, giving the expression of all *characteristic classes* on a manifold as integrals of polynomials in the curvature of any connection. (For a systematic account, cf. [95].) These integrals interact in a very interesting way with some variational problems. We saw in particular how a topological lower bound of the Yang-Mills functional on oriented 4-dimensional manifolds is crucial in detecting the self-dual and anti-self-dual connections. The same happens also for the energy of maps between Riemann surfaces.

Quite often these integrals also provide topological obstructions to the existence of critical points of some variational problems. This is the case for example for the Thorpe-Hitchin obstruction to the existence of Einstein metrics on compact 4-dimensional manifolds (cf. [21]).

2.4. Chern-Weil theory can be refined in the Kählerian setting, because one works there most often in a given Kähler class. Therefore more integrals can be introduced which have a purely cohomological : besides the ones corresponding to Chern numbers, one can introduce Kähler invariants which mix the (fixed) Kähler class and the Chern classes.

2.5. This can be refined in the use of *secondary* characteristic classes. An obstruction to the existence of solutions to a variational problem is given by Futaki's invariant, which can be thought as such a class, cf. [66]. These classes depend on the choice of a connection, hence can be used as Lagrangians. This was for example done by S. Donaldson in [41] and developed further by T. Mabuchi in [90] and [91].

3. Interactions with Physics

The interplay of Mathematics with Physics has been acknowledged all along the history of both fields. The most classical texts on this matter may be due to H. Weyl (cf. [143] for example) and E.P. Wigner (cf. [144]). Nevertheless, the last decade has seen such an overwhelming flow of notions, problems and suggestions coming from Physics that a specific sub-section has to be devoted to this aspect (for an interesting account, cf. [147]). In the Natural Sciences, it has become a tradition to tie the Laws of Nature to Variational Pricniples. This goes back to Descartes, Huyghens, Euler, Lagrange, Maupertuis.

A) Classical Field Theory

3.1. Both Electromagnetism and General Relativity are examples of physical theories which can be tied in a very direct way to important variational problems that we discussed in this survey. For General Relativity, this goes back explicitly to Hilbert, cf. [72]. But both of these theories have developed along this century. The first one has given rise to *gauge field theories*, in particular Yang-Mills theory, and for the second a notable attempt to generalize it is *supergravity* (for a reference, cf.[50]).

3.2. It is also worth noticing that even in its classical aspects General Relativity has been enriched by new points of view, such as the ADM formulation of the Einstein equations (cf. [8] and [62]) or the concept of black holes. (For a modern survey, [59] is still a good reference). This has naturally lead to a rather deep study of isolated systems, modelled by asymptotically flat spaces which provide a good setting for studying interesting problems on non compact manifolds (cf. for example [17] and [56]).

B) New Challenges coming from Quantum Theory

3.3.   The most drastic change in XXth century Physics has undoubtly come from Quantum Mechanics. This has forced a complete reformulation of the fundamental laws of Physics. There is also a mathematical price to it with the appearance of very sophisticated mathematical objects, some of them as path integrals or Feynmann integrals not yet defined in a mathematically rigorous way (for formulations connected to this, cf. [1]). Attempts have of course been made to connect this to more classical formulations (cf. for example [76]).

3.4.   The study of many variational problems have been greatly stimulated by modern Physics, in particular *string theory*. (To quote a few very important papers, cf. [111], [112], [122], [148], [149], [150],)

3.5.   This has lead mainly under the vision of E. Witten to very interesting new viewpoints on topology of low dimensional manifolds by suggesting the possibility some highly non-trivial invariants in a way which was connected to variational problems, cf. [10], [13] or [11].

C) Other Physical Theories

3.6.   This extremely rich harvest of new results and phenomenon through a quantum look at some mathematical problems should not make us forget the deep connection of eigenvalue problems with physical theories. (For a historical account of this, cf. [110].)

3.7.   In recent years, the most interesting stimulations came from the theory of liquid crystals (cf. [25]) and that of membranes (cf. [3]).

Bibliography

[1] ALBEVERIO, S., PAYCHA, S., SCARLATTI, S. *A Short Overview of Mathematical Approaches to Functional Integration*, in *Functional Integration, Geometry and Strings* (ed. Z. Haba, J. Sobczyk), Prog. in Phys. **13** (1989), Birkhäuser, Basel,230–276.

[2] ALLARD, W.K., *On the First Variation of a Varifold : Boundary Behaviour*, Ann. Math. **101** (1975), 418–446.

[3] ANDERSON, D., HOFFMAN, D., HENKE, C., THOMAS, E.L., *Periodic Area-Minimizing Surfaces in Block Copolymers*, Nature **334** (1988), 598–601.

[4] ANDERSON, M., *Ricci Curvature Bounds and Einstein Metrics on Compact Manifolds*, J. Amer. Math. Soc. **2** (1990), 455–490.

[5] ANDERSON, M., *The L^2-Structure of Moduli Spaces of Einstein Metrics on 4-Manifolds*, J. Geom. Functional Anal. **2** (1992), 29–89.

[6] ANDERSON, M., *Degeneration of Metrics with Bounded Curvature and Applications to Critical Metrics of Riemannian Functionals*, in *Differential Geometry* (ed. R. Greene, S.T. Yau), Proc. Amer. Math. Soc. Symp. Pure Math. **54** (1993), 53–79.

[7] ANDERSSON, S., HYDE, S.T., LARSSON, K., LIDIN, S., *Minimal Surfaces and Structures : From Inorganic and Metal Crystals to Cell Membranes and Biopolymers*, Chem. Rev. **88** (1988), 221–242.

[8] ARNOWITT, R., DESER, S., MISNER, C., *Coordinate Invariance and Energy Expressions in General Relativity*, Phys. Rev. **122** (1961), 997–1006.

[9] ATIYAH, M.F., *Geometry of Yang-Mills Fields*, Lezioni Fermiane, Acad. Naz. dei Lincei, Pisa, 1979.

[10] ATIYAH, M.F., *New Invariants of 3- and 4-Dimensional Manifolds*, in *Symposium on the Mathematical Heritage of Hermann Weyl* (ed. R.O. Wells), Amer. Math. Soc., Providence, 1988, 285–299.

[11] ATIYAH, M.F., *Topological Quantum Field Theories*, Publ. Math. Inst. Hautes Etudes Sci. **68** (1989), 175–186.

[12] ATIYAH, M.F., BOTT, R., *The Yang-Mills equations over a Riemann Surface*, Sir Michael Atiyah Collected Works, Oxofrd Univ. Press, Oxford.

[13] ATIYAH, M.F., JEFFREY, L., *Topological Lagrangians and Cohomology*, J. Geom. Phys. **7** (1991), 119–136.

[14] AUBIN, T., *Equations différentielles non linéaires et problème de Yamabe concernant la courbure scalaire*, J. Math. Pures Appl. **55** (1976), 269–296.

[15] BANDO, S., *Bubbling out of Einstein Manifolds*, Tôhoku Math. J. **42** (1990), 205–216.

[16] BANDO, S., KASUE, A., NAKAJIMA, H., *On a Construction of Coordinates at Infinity on Manifolds with Fast Curvature Decay and Maximal Volume Growth*, Inventiones Math. **97** (1989), 313–349.

[17] BARTNIK, R., *The Mass of an Asymptotically Flat Manifold*, Commun. Math. Phys. **39** (1986), 661–693.

[18] BERGER, M., *Sur les premières valeurs propres des variétés riemanniennes*, Compositio Math. **26** (1973), 129–149.

[19] BERGER, M., GAUDUCHON, P., MAZET, E., *Le spectre d'une variété riemannienne*, Lect. Notes in Math. **194**, Springer-Verlag,Berlin-Heidelberg, 1971,

[20] BERGER, M., LASCOUX, A., *Variétés kählériennes compactes*, Lect. Notes in Math. **154**, Springer-Verlag, Berlin-Heidelberg, 1970.

[21] BESSE, A.L., *Einstein Manifolds*, Erg. Math. Grenzgebiete **10**, Springer-Verlag, Heidelberg-Berlin-New York, 1987.

[22] BESSON, G., COURTOIS, G., GALLOT, S., *Volume minimal des espaces localement symétriques*, Inventiones Math. **103** (1991), 417–445.

[23] BESSON, G., COURTOIS, G., GALLOT, S., *Les variétés hyperboliques sont des minima locaux de l'entropie topologique*, Inventiones Math., (to appear).

[24] BÉTHUEL, F., *The approximation problem for Sobolev mappings between manifolds*, Acta Math. **167** (1991).

[25] BÉTHUEL, F., BRÉZIS, H., CORON, J.M., *Relaxed Energies for Harmonic Maps*, in *Variational Methods* (ed. H. Berestycki,J.M. Coron, I. Ekeland), Birkhäuser, Basel, 1990.

[26] BOMBIERI, E., DE GIORGI, E., GIUSTI, E., *Minimal Cones and Bernstein Problem*, Inventionès Math. **7** (1968), 243–268.

[27] BOURGUIGNON, J.-P., *Analytical Problems Arising in Geometry : Examples from Yang-Mills Theory*, Jahresber. der Deutschen Math. Ver. **87** (1985), 67–89.

[28] BOURGUIGNON, J.-P., GAUDUCHON, P., *Spineurs, opérateurs de Dirac et variations de métriques*, Commun. Math. Phys. **144** (1992), 581–599.

[29] BOURGUIGNON, J.-P,. LAWSON, H.B., *Stability and Isolation Phenomena for Yang-Mills Fields*, Commun. Math. Phys. **79** (1981), 189–230.

[30] BOURGUIGNON, J.-P,. LAWSON, H.B., *Yang-Mills Theory : its Physical Origin and Differential Geometric Aspects*, in *Seminar on Differential Geometry* (ed. S.T. Yau), Ann. Math. Studies **102**, Princeton Univ. Press, Princeton, 1982, 395–421.

[31] BOURGUIGNON, J.-P., LI, P., YAU, S.T., *Upper Bound for the First Eigenvalue of Algebraic Submanifolds,*, Comment. Math. Helvetici 69(1994), 199–207.

[32] BRYANT, R., *A Duality Theorem for Willmore Surfaces*, J. Differential Geom. **20** (1984), 23–53.

[33] CALABI, E., *Minimal Immersions of Surfaces into Euclidean Spheres*, J. Differential Geom. **1** (1967), 111–125.

[34] CALABI, E., *Extremal Kähler Metrics*, in *Seminar on Differential Geometry* (ed. S.T. Yau), Ann. Math. Studies **102**, Princeton Univ. Press, Princeton, 1982, 259–290.

[35] CALABI, E., *Extremal Kähler Metrics II*, in *Differential Geometry and Complex Analysis* (ed. I. Chavel and H.M. Farkas),Springer-Verlag, New York, 1985.

[36] CHEEGER, J., EBIN, D., *Comparison Theorems in Riemannian Geometry*, North-Holland, Amsterdam, 1974.

[37] CHEN, Y., DING, W.Y., *Blow-up and Global Existence of Heat Flow of Harmonic Maps*, Inventiones Math. **99** (1990), 567–578.

[38] CONCUS, P., FINN, R. *Variational Methods for Free Surface Interfaces*, Springer-Verlag, Berlin-Heidelberg-New York, 1987.

[39] CONNES, A., *Non-Commutative Geometry*, to appear.

[40] DONALDSON, S.K., *An Application of Gauge Theory to the Topology of 4-Manifolds*, J. Differential Geom. **18** (1983), 279–315.

[41] DONALDSON, S.K., *Anti-Self-Dual Yang-Mills Connections over Complex Algebraic Surfaces and Stable Vector Bundles*, Proc. London Math. Soc. **50** (1983), 1–26.

[42] DONALDSON, S.K., *La topologie différentielle des surfaces complexes*, C. R. Acad. Sci. Paris **301** (1985), 317–320.

[43] DONALDSON, S.K., *Connections, Cohomology and the Intersection Form of 4-Manifolds*, J. Differential Geom. **24** (1986), 275–342.

[44] DONALDSON, S.K., *The Orientation of Yang-Mills Moduli Spaces and 4-Manifold Topology*, J. Differential Geom. **26** (1987), 141–168.

[45] DONALDSON, S.K., *Infinite Determinants, Stable Bundles, and Curvature*, Duke Math. J. **54** (1987), 231–247.

[46] DONALDSON, S.K., *Polynomial Invariants for Smooth Four-Manifolds*, Topology **29** (1990), 257–315.

[47] DONALDSON, S.K., *Yang-Mills Invariants of Four-Manifolds*, in *Geometry of Low-Dimensional Manifolds*, London Math. Soc. Lect. Notes Series **150**, Cambridge Univ. Press, Cambridge, 1990, 5–40.

[48] DONALDSON, S.K., KRONHEIMER, P., *The Geometry of Four-Manifolds*, Oxford Univ. Press, Oxford, 1990.

[49] DOUGLAS, J., *Solution to the Problem of Plateau*, Trans. Amer. Math. Soc. **33** (1931), 263–321.

[50] DUFF, M.J., NILSSON, B.E.W., POPE, C.N., *Kaluza-Klein Supergravity*, Phys. Rep. **130** (1986), 1-142.

[51] EELLS, J., *A Setting for Global Analysis*, Bull. Amer. Math. Soc. **72** (1966), 751–807.

[52] EELLS, J., LEMAIRE, L., *A Report on Harmonic Maps*, Bull. London Math. Soc. **10** (1978), 1–68.

[53] EELLS, J., LEMAIRE, L., *Selected Topics on Harmonic Maps*, C.B.M.S. Regional Conf. Ser. **50**, Amer. Math. Soc., Providence, 1983.

[54] EELLS, J., LEMAIRE, L., *Another Report on Harmonic Maps*, Bull. London Math. Soc. **20** (1988), 385–524.

[55] EELLS, J., SAMPSON, J., *Harmonic Mappings of Riemannian Manifolds*, Amer. J. Math. **86** (1964), 109–160.

[56] EGUCHI, T., HANSON, A., *Asymptotically Flat Solutions to Euclidean Gravity*, Phys. Lett. **B 74** (1978), 249–251.

[57] EINSTEIN, A., *Die Grundlage der Allgemeinen Relativitätstheorie*, Ann.Phys. **49** (1916), 769–822.

[58] EINSTEIN, A., GROSSMANN, M., *Entwurf einer Allgemeinerten Relativitätstheorie und einer Theorie des Gravitations I : Physikalischer Teil*, Z. Math. Phys. **62** (1913), 225–244 ;*idem II : Mathematischer Teil*, ibidem, 244–261.

[59] ELLIS, G.F.R., HAWKING, S.W., *The Large Scale Structure of Space-Time*, Cambridge Univ. Press, Cambridge, 1973.

[60] FEDERER, H., *Geometric Measure Theory*, Springer-Verlag, New York, 1969.

[61] FEYNMAN, R.P., *QED : A Critical Path between Light and Matter*, Penguin Books, Penguin

[62] FISCHER, A.E., MARSDEN, J.E., *The Einstein Equations of Evolution : a Geometric Approach*, J. Math. Phys. **13** (1972),546–568.

[63] FLOER, A., *An Instanton Invariant for 3-Manifolds*, Commun. Math. Phys. **118** (1988), 215–240.

[64] FREED, D.S., UHLENBECK, K.K., *Instantons and Four-Manifolds*, Math. Sci. Res. Inst. Publications 1 (1984), Springer,Berlin-Heidelberg-New York.

[65] FREEDMAN, M.H., HE, Z.X., WANG, Z., *On the Energy of Knots and Unknots*, Preprint, 1993.

[66] FUTAKI, A., *An obstruction to the existence of Kähler-Einstein metrics*, Inventiones Math. **73** (1983), 437–443.

[67] GAO, L., *Einstein Manifolds*, J. Differential Geom. **32** (1990), 155–183.

[68] GALLOT, S., HULIN, D., LAFONTAINE, J., *Riemannian Geometry*, Grad. Texts. in Math., Springer-Verlag, Berlin-Heidelberg-New York, 1992,

[69] HARDT, R., LIN, F.H., POON, C., *Axially Symmetric Harmonic Maps Minimizing a Relaxed Energy*, Commun. Pure Appl. Math. **45** (1992), 417–459.

[70] HERSCH, J., *Caractérisation variationnelle d'une somme de valeurs propres consécutives*, C.R. Acad. Sci. Paris **252** (1961), 1714–1716.

[71] HERSCH, J., *Quatre propriétés isopérimétriques de membranes sphériques homogènes*, C.R. Acad. Sci. Paris **270** (1970),1645–1648.

[72] HILBERT, D. *Die Grundlagen der Physik (Erste Mitteilung)*, Nachr. Gesellsch. Wiss. Göttingen **3** (1915), 395–407.

[73] HSIANG, W.Y., *Minimal Cones and the Spherical Bernstein Problem, I*, Ann. Math. **118** (1983), 61–73 ; *idem, II*, Inventiones Math. **74** (1983), 351–369.

[74] ITOH, M., *Invariant Connections and Yang-Mills Solutions*, Trans. Amer. Math. Soc. **267** (1981), 229–236.

[75] JAFFE, A., TAUBES, C.H., *Vortices and Monopoles*, Progress in Phys. **2**, Birkhäuser, Boston (1980).

[76] JACKIW, R., *Quantum Meaning of Classical Field Theory*, Rev. Mod. Phys. **49** (1977), 681–706.

[77] JENSEN, G., *The Scalar Curvature of Left-Invariant Riemannian Metrics*, Indiana Univ. Math. J. **20** (1971), 715–737.

[78] JOST, J., PENG, X.W., *Group Actions, Gauge Transformations, and the Calculus of Variations*, Math. Ann. **293** (1992), 595–621.

[79] KARCHER, H., *Construction of Minimal Surfaces*, Suveys in Geometry, Tokyo Univ. (1989), 1–96.

[80] KAZDAN, J.L., WARNER, F.W., *Prescribing Curvatures*, in *Differential Geometry*, Proc. Amer. Math. Soc. Symp. Pure Math. **27** (1975), 309–319.

[81] KIM, D., KUSNER, R., *Torus Knots Extremizing the Conformal Energy*, Preprint, Univ. Mass. Amherst, 1993.

[82] KOISO, N., *Rigidity and Infinitesimal Deformability of Einstein Metrics*, Osaka J. Math. **29** (1982), 643–668.

[83] LAGRANGE, J.L., *Mechanique Analitique*, 1788.

[84] LANCZOS, C., *A Remarkable Property of the Riemann-Christoffel Tensor in Four Dimensions*, Ann. Math. **39** (1938), 842–850.

[85] LANGER, J., SINGER, D.A., *Knotted Elastic Curves in* $\mathbf{R}^3$, J. London Math. Soc. **30** (1984), 512–520.

[86] LAWSON, H.B., *Minimal Varieties*, in *Differential Geometry*, Proc. Amer. Math. Soc. Symp. Pure Math. **27** (1975), 143–175.

[87] LAWSON, H.B., *The Theory of Gauge Fields in Four Dimensions*, C.B.M.S. Regional Conference Series **58**, 1985.

[88] LEE, J., PARKER, T., *The Yamabe problem*, Bull. Amer. Math. Soc. **17** (1987), 37–91.

[89] LI, P., YAU, S.T., *A New Conformal Invariant and its Application to the Willmore Conjecture and the First Eigenvalue of Compact Surfaces*, Inventiones Math. **69** (1982), 269–291.

[90] MABUCHI, T., *A Functional Integrating Futaki's Invariant*, Proc. Japan Acad. Sci. **58** (1985), 119–120.

[91] MABUCHI, T., *Einstein-Kähler forms, Futaki Invariants and Convex Geometriy on Toric Fano Varieties*, Osaka J. Math. **24** (1987), 705–737.

[92] MAZET, E., *La formule de la variation seconde de l'énergie au voisinage d'une application 'harmonique*, J. Differential Geom. **8** (1973), 279–296.

[93] MEEKS, W.H., YAU, S.T., *The Classical Plateau Problem and the Topology of three-Dimensional Manifolds*, Topology **21** (1982), 409–442.

[94] MILNOR, J.W., *Morse Theory*, Ann. Math. Studies **51**, Princeton Univ. Press, Princeton, 1963.

[95] MILNOR, J.W., STASHEFF, J., *Characteristic classes*, Ann. Math. Studies **76**, Princeton Univ. Press, Princeton, 1974.

[96] MORSE, M,. *The Calculus of Variations in the Large*, Princeton Univ. Press, Princeton.

[97] NAKAJIMA, H., *Hausdorff Convergence of Einstein 4-Manifolds*, J. Fac. Sci. Univ. Tokyo **35** (1988), 411–424.

[98] NAKAJIMA, H., *Moduli spaces of Anti-Self-Dual Connections on ALE Gravitational Instantons*, Inventiones Math. **102** (1990), 267–303.

[99] OHNITA, Y., *Stability of Harmonic Maps and Standard Minimal Immersions*, Tôhoku Math. J. **38** (1986), 259–267.

[100] OSGOOD, B., PHILLIPS, R., SARNAK, P., *Extremals of Determinants of Laplacians*, J. Functional Anal. **80** (1988), 148–211.

[101] OSGOOD, B., PHILLIPS, R., SARNAK, P., *Compact Isospectral Sets of Surfaces*, J. Functional Anal. **80** (1988), 212–234.

[102] OSGOOD, B., PHILLIPS, R., SARNAK, P., *Moduli Space, Heights and Isospectral Sets of Plane Domains*, Ann. Math. **129** (1989), 293–362.

[103] OSSERMAN, R., *The Isoperimetric Inequality*, Bull. Amer. Math. Soc. **84** (1978), 1183-1238.

[104] PALAIS, R.S., *Global Non-Linear Analysis*, Benjamin, New York, 1964.

[105] PALAIS, R.S., *Lusternik-Schnirelman Theory in Banach manifolds*, Topology **12**, (1972).

[106] PALAIS, R.S., *The Principle of Symmetric Criticality*, Commun. Math. Phys. **145** (1979), 19–30.

[107] PALAIS, R.S., SMALE, S., *A generalized Morse theory*, Bull. Amer. Math. Soc. **70** (1964), 165–172.

[108] PARKER, T., *Non Minimal Yang Mills Fields and Dynamics*, Invent. Math. **107** (1992), 397–402.

[109] PLATEAU, J.A.F., *Statique expérimentale et théorique des liquides soumis aux seules forces moléculaires*, Gauthier-Villars, Paris, 1873.

[110] POLYA, G., SZEGÖ, G., *Isoperimetric Inequalities in Mathematical Physics*, Ann. Math. Studies **27**, Princeton Univ. Press, Princeton, 1951.

[111] POLYAKOV, A., *Quantum Geometry of Bosonic Strings*, Phys. Lett. **B 103** (1981), 207-210.

[112] POLYAKOV, A., *Quantum Geometry of Fermionic Strings*, Phys. Lett. **B 103** (1981), 211-213.

[113] RADO, T., *On Plateau's Problem*, Ann. Math. **31** (1930), 457–469.

[114] SACKS, J., UHLENBECK, K.K., *The Existence of Minimal Immersions of 2-Spheres*, Ann. Math. **113** (1981), 1–24.

[115] SADUN, L., SEGERT, J., *Non-Self-Dual Yang-Mills Connections with Quadruple Symmetry*, Commun. Math. Phys. **145** (1992), 363–391.

[116] SCHOEN, R., *Conformal Deformation of a Riemannian Metric to Constant Curvature*, J. Differential Geom. **20** (1984), 479–495.

[117] SCHOEN, R., *Variational Theory for the Total Scalar Curvature Functional for Riemannian Metrics and Related Topics*, in *Topics in Calculus of Variations, Seminar 1987* (ed. M. Giaquinta), Lect. Notes in Math. **1365**, Springer-Verlag, New York, 1989, 120–154.

[118] SCHOEN, R., UHLENBECK, K.K., *A Regularity Theory of Harmonic Maps*, J. Differential Geom. **17** (1982), 307-335.

[119] SCHOEN, R., YAU, S.T., *On the Proof of the Positive Mass Conjecture I*, Commun. Math. Phys. **65** (1978), 45–76.

[120] SCHOEN, R., YAU, S.T., *The Existence of Incompressible Minimal Surfaces and the Topology of Three-Dimensional Manifolds with Nonnegative Scalar Curvature*, Ann. Math. **110** (1979), 127–142.

[121] SCHOEN, R., YAU, S.T., *On the Proof of the Positive Mass Conjecture II*, Commun. Math. Phys. **79** (1981), 231–260.

[122] SEGAL, G.B., *Loop Groups and Harmonic Maps*, in *Advances in Homotopy Theory*, London Math. Soc. Lect. Notes **139**, Cambridge Univ. Press, Cambridge, 1989, 153–164.

[123] SÉMINAIRE PALAISEAU, *Théorie des variétés minimales et applications*, Astérisque **154–155** (1987).

[124] SIBNER, L.M., SIBNER, R.J., UHLENBECK, K.K., *Solutions to Yang-Mills Equations which are Non-Self-Dual*, Proc. Nat. Acad. Sci. U.S.A. **86** (1989), 8610-8613.

[125] SIMON, B., *Functional Integration and Quantum Physics*, Academic Press, New York, 1979.

[126] SIMON, L., *Survey Lectures on Minimal Submanifolds*, in *Seminar on Minimal Submanifolds*, E. Bombieri ed., Ann. Math. Studies **103**, Princeton Univ. Press, Princeton, 1983.

[127] SIMONS, J., *Minimal Varieties in Riemannian Manifolds*, Ann. Math. **88** (1968), 62–105.

[128] SMITH, R.T., *The Second Variation Formula for Harmonic Maps*, Proc. Amer. Math. Soc. **47** (1975), 229-236.

[129] STRUWE, M., *On the Evolution of Harmonic Maps of Riemannian Surfaces*, Comment. Math. Helvetici **60** (1985), 558-581.

[130] TAUBES, C.H., *Self-Dual Yang-Mills Connections on Non Self-Dual 4-Manifolds*, J. Differential Geom. **17** (1982), 139–170.

[131] TAUBES, C.H., *Casson's Invariant and Gauge Theory*, J. Differential Geom. **31** (1990), 547–599.

[132] TAYLOR, J., *The Structure of Singularities in Soap-Bubble-Like and Soap-Film-Like Minimal Surfaces*, Ann. Math. **103** (1976), 489–539.

[133] TRUDINGER, N., *Remarks Concerning the Conformal Deformation of Riemannian Structures on Compact Manifolds*, Ann. Scuola Norm. Sup. Pisa **22** (1968), 265–274.

[134] UHLENBECK, K.K., *Removable Singularities in Yang-Mills fields*, Commun. Math. Phys. **83** (1982), 11–29.

[135] UHLENBECK, K.K., *Variational Problems for Gauge Fields*, Proc. Int. Cong. Math. (1982), Warsaw, 585–591.

[136] UHLENBECK, K.K., *Connections with L^p-Bounds on Curvature*, Commun. Math. Phys. **83** (1982), 31–42.

[137] UHLENBECK, K.K., *Harmonic Maps into Lie Groups (Classical Solutions of the Chiral Model)*, J. Differential Geom. **30** (1989), 1–50.

[138] URAKAWA, H., *Equivariant Theory of Yang-Mills Connections over Riemannian Manifolds of Cohomogeneity One*, Indiana Univ. Math. J. **37**(1988), 753–788.

[139] VOLTERRA, R., *Leçons sur le Calcul des Variations*, 1898.

[140] WARNER, F.W., *Foundations of Differentiable Manifolds and Lie groups*, (second edition), Grad. Texts in Math., Springer-Verlag, Berlin-Heidelberg-New York,1983.

[141] WENTE, H. *A Counterexample to a Conjecture of H. Hopf*, Pacific. J. Math. **121** (1986), 193-244.

[142] WEYL, H., *Raum, Zeit, Materie*, (1920) ; *Space, Time, Matter*, Engl. version, Dover, New York, (1952).

[143] WEYL, H., *Philosophy of Mathematics and Natural Science*, (transl. from *Philosophie der Mathematik und Naturwissenschaft* in *Handbuch der Philosophie*, ed. R. Oldenburg), Princeton Univ. Press, Princeton, 1949.

[144] WIGNER, E.P., *The Unreasonable Effectiveness of Mathematics in the Natural Sciences*, Commun. Pure Applied Math. **13** (1960), 1–14.

[145] WILLMORE, T., *Note on Embedded Surfaces*, Anal. Stüntifice ale Univ. Iasi, Sect. I Mat. **11** (1965), 493–496.

[146] WITTEN, E., *Supersymmetry and Morse Theory*, J. Differential Geom. **17** (1982), 661–692.

[147] WITTEN, E., *Physics and Geometry*, Proc. Int. Congr. Math. Berkeley (1981), 267–303.

[148] WITTEN, E., *Topological Sigma-Models*, Commun. Math. Phys. **118** (1988), 411–449.

[149] WITTEN, E., *Quantum Field Theory and the Jones Polynomial*, Commun. Math. Phys. **121** (1989), 351–400.

[150] WITTEN, E., *Two-Dimensional Gravity and Intersection Theory on Moduli Space*, Surveys in Differential Geom. **1** (1991), 243–310.

[151] YAMABE, H., *On a Deformation of Riemannian Structures on Compact Manifolds*, Osaka Math. J. **12** (1960), 21–37.

Geometry of Gauge Fields

Kenji Fukaya*

Department of Mathematical Sciences
University of Tokyo
Hongô, Bunkyô-ku, Tokyo, 113
Japan

This article is an extended version of the union of two series of lectures delivered by the author in 1993 July. One is the lecture entitled "Geometry of Gauge Fields" at the first MSJ International Research Institute on Geometry and Global Analysys, held at Tohoku University, Japan and the other is those entitled as "Gauge theory and Topological Field Theories" at "The 13-th DaeWoo Workshop on Pure Mathematics" held at Pohang National University, Korea. The purpose of those lectures are to illustrate a rough account of gauge theory and its applications to low dimensional topology without going into technical details. For this purpose no effort was made toward the accuracy or preciseness of the contents. Instead the author try to explain ideas of "infinite dimensional geometry" which are sometimes hidden under rigorous mathematical arguments in the literature.

Needless to say, this article is far from a comprehensive survey of the works in gauge theory. Especially, one of the most important recent progress due to Kronheimer and Mrowka using singular connections is not discussed. Also there is no explicit calculation of the invariant and hence no explicit application to low dimensional topology. Moreover the author confines himself to those topics related to Yang Mills equations, though the scope of gauge theory is much broader than Yang Mills theory. Especially there is no description on the new exiting progress based on Seiberg-Witten equation, since this is a record of the lecture given before Seiberg-Witten equation was discovered.

The author would like to thank the organizers of the two symposiums, especially to Professors Hongjong Kim and Seiki Nishikawa.

* Current Address: Department of Mathematics, Faculty of Science, Kyoto University, Kitasirakawa, Sakyo-ku, Kyoto, 606 Japan

§ 1 Donaldson Invariant of 4 manifolds

First, let us recall some classical ideas which have been used since 1950's, the very beginning of differential topology. Suppose we have two smooth manifolds M_1 and M_2. Assume that they are homotopy equivalent. How can we show that they are not diffeomorphic?

One can use algebraic topology for this purpose but not the algebraic topology of M_1, M_2 itself. (They have the same invariant.) Instead we can use the fact that the smooth structure automatically induces a vector bundle, that is, the tangent bundle TM_i. So we take characteristic classes of tangent bundles as invariants of smooth manifolds M_i. These invariants lie in the cohomology groups of the manifolds. (Those cohomology groups are isomorphic to each other.) This idea and its modification were used in 1950's to find various exotic smooth structures. Subsequently D. Sullivan proved that characteristic classes (of tangent bundles) determine the differentiable structures of a manifold up to finite ambiguity in dimension ≥ 5.[2]

In dimension four, we now know that the characteristic classes of tangent bundles are not sufficient to determine the differentiable structures. ([D5])

So what can we use instead ? Donaldson invariants ([D4]) are obtained, roughly speaking, by considering the characteristic classes of an infinite dimensional bundle over the infinite dimensional manifold which we can associate in a canonical way to a 4-manifold.

Now, let us take a 4-manifold M. We first associate an infinite dimensional manifold to our manifold M as follows. Let G be a compact Lie group and $P \to M$ be a principal G-bundle over M. We take the set of all G-connections of this bundle. Let us call it $\mathcal{A}(M,P)$. Since the bundle P is determined completely by its characteristic classes, it follows that we can find "the same" bundle over M' if M' is homotopy equivalent to M. But $\mathcal{A}(M,P)$ is too simple for our purpose, since it is an affine space and hence there is no interesting algebraic topology on it. So, to get something nontrivial, we need to divide it by the group of gauge transformations $\mathcal{G}(M,P)$. Here $\mathcal{G}(M,P)$ denotes the set of all bundle isomorphisms which cover the identity map of M. In other words,

$$\mathcal{G}(M,P) = \left\{ \varphi : P \to P \left| \begin{array}{l} \pi f = \pi \\ \varphi(xg) = \varphi(x)g \text{ for } \forall x \in P, g \in G \end{array} \right. \right\}.$$

---

[2]One need to assume that manifold is simply connected.

Here $\pi: P \to M$ is the projection of fibre bundle and we regard P as a space on which G has a free right action.

The group $\mathcal{G}(M, P)$ acts on $\mathcal{A}(M, P)$ by

$$\varphi^*(\nabla)(s) = \varphi^{-1} \circ \nabla \circ \varphi(s).$$

Let $\mathcal{B}(M, P)$ denote the quotient space of this action. Let me first remark :

Fact 1.1.   The homotopy type of $\mathcal{B}(M, P)$ depends only on the homotopy type of M (and the Lie group G).

Now, we want to find a vector bundle E on $\mathcal{B}(M, P)$. Roughly speaking, the Poincaré dual of the Euler class of E is the invariant we want to define here. Since the "manifold" $\mathcal{B}(M, P)$ is of infinite dimension, one needs to take an infinite dimensional vector bundle in order to get a finite dimensional class as a Poincaré dual of its Euler class.

We count the dimension of $\mathcal{B}(M, P)$ as follows. Let g be the Lie algebra of G and g be its dimension. Then the connection is locally a g-valued one form. Therefore the dimension of $\mathcal{A}(M, P)$ is roughly $4g$ times the dimension of the function space over M. Next, obviously the dimension of $\mathcal{G}(M, P)$ is g times the dimension of the function space over M. So the dimension $\mathcal{B}(M, P)$ is $3g$ times the dimension of the function space over M.

So we need to find a vector bundle over $\mathcal{B}(M, P)$ whose dimension is 3 times the dimension of the function space over M. Such a vector bundle is obtained in the following way. We take the adjoint representation $ad: G \to \text{Aut}(g)$ and then get a vector bundle $AdP = P \times_{ad} g$, that is the quotient of the direct product $P \times g$ by the diagonal action of G. We next introduce a Riemannian metric on M. It induces the Hodge $*$-operator on the set of differential forms on M. We put

$$\Lambda^2_+ = \left\{ u \in \Lambda^2 \mid *u = u \right\}.$$

Define Λ^2_- in a similar way.

Let p be a point in M and $e^i, i = 1, \cdots, 4$ be an orthonormal frame of the cotangent bundle in a neighborhood of p in M. Then a frame of Λ^2_+ is given by $e^1 \wedge e^2 + e^3 \wedge e^4$, $e^1 \wedge e^3 - e^2 \wedge e^4$, $e^1 \wedge e^4 + e^2 \wedge e^3$. Hence the dimension of the set of all sections of the bundle $\Lambda^2_+ \otimes AdP$ is $3g$ times one of the function space over M, the one we are looking for.

We then construct an (infinite dimensional) vector bundle over $\mathcal{B}(M,P)$ as follows. We remark that the group $\mathcal{G}(M,P)$ acts on $\Gamma(\Lambda_+^2 \otimes \mathrm{Ad}P)$ (the set of smooth sections) from the left and on $\mathcal{A}(M,P)$ from the right. Then we take the quotient $\mathcal{A}(M,P) \times_{\mathcal{G}(M,P)} \Gamma(\Lambda_+^2 \otimes \mathrm{Ad}P)$ of $\mathcal{A}(M,P) \times \Gamma(\Lambda_+^2 \otimes \mathrm{Ad}P)$ by the diagonal action. Thus we have a vector bundle

$$\mathcal{A}(M,P) \times_{\mathcal{G}(M,P)} \Gamma(\Lambda_+^2 \otimes \mathrm{Ad}P) \to \mathcal{B}(M,P)$$

"Definition" 1.2.   We call the Euler class

$$e(\mathcal{A}(M,P) \times_{\mathcal{G}(M,P)} \Gamma(\Lambda_+^2 \otimes \mathrm{Ad}P)) \in H^{\infty-k}(\mathcal{B}(M,P);\mathbf{Z}) \approx H_k(\mathcal{B}(M,P);\mathbf{Z})$$

of this bundle to be the *Donaldson invariant* of M. (Here k is a finite number and $H^{\infty-k}(\mathcal{B}(M,P);\mathbf{Z}) \approx H_k(\mathcal{B}(M,P);\mathbf{Z})$ is the Poincaré duality.)

Since the fibre of the vector bundle $\mathcal{A}(M,P) \times_{\mathcal{G}(M,P)} \Gamma(\Lambda_+^2 \otimes \mathrm{Ad}P) \to \mathcal{B}(M,P)$ is infinite dimensional, 1.2 is not a rigorous definition. Moreover, since the infinite dimensional orthogonal group is contractible, it is not easy to define characteristic classes of infinite dimensional vector bundles. To give a general framework for such characteristic classes is still quite difficult open problem. Instead of defining it in general, we study a specific example and try to define $e(\mathcal{A}(M,P) \times_{\mathcal{G}(M,P)} \Gamma(\Lambda_+^2 \otimes \mathrm{Ad}P))$.

In general if one has a vector bundle of finite rank $E \to X$ over a manifold of finite dimension, X, then the Poincaré dual of its Euler class $e(E)$ is given as follows : let $s:X \to E$ be a section which is transversal to the zero section. We have :

$$PD(e(E)) = [s^{-1}(0)].$$

Here we give an orientation on $s^{-1}(0)$ using the ones on E and X.

Now let us go back to the infinite dimensional situation. We need to find a section of the bundle $\mathcal{A}(M,P) \times_{\mathcal{G}(M,P)} \Gamma(\Lambda_+^2 \otimes \mathrm{Ad}P) \to \mathcal{B}(M,P)$. There is an obvious natural choice for it. Namely, we put

$$s([A]) = [(A, F_A + {}^*F_A)].$$

Here F_A is the curvature of A and $*$ is the Hodge $*$ operator. Hereafter we write $F_A^+ = \dfrac{F_A + *F_A}{2}$. Hence

$$PD(e(\mathcal{A}(M,P) \times_{\mathcal{G}(M,P)} \Gamma(\Lambda_+^2 \otimes \mathrm{Ad}P))) = [\mathcal{M}(M,P)],$$

where

$$\mathcal{M}(M,P) = \left\{ [A] \in \mathcal{B}(M,P) \,\middle|\, F_A^+ = 0 \right\}.$$

We call an element of $\mathcal{M}(M,P)$ an *ASD (anti-self-dual) connection*. Thus we find that the cycle of $\mathcal{B}(M,P)$ represented by the moduli space of ASD connections is regarded as the Poincaré dual of an (infinite dimensional) vector bundle.

We remark that, in our situation, the homology class $[s^{-1}(0)]$ *does* depend on the choice of the section s, (which we took $s(A) = F_A^+$). But (roughly speaking) it is independent of the deformations of it in the following two categories. Let s_1, s_2 be two sections.

(1)   There exists a homotopy s_t, $t \in [1,2]$, joining s_1 to s_2 such that $s_t^{-1}(0)$ is of finite dimension for each t.

(2) $s_1 - s_2$ is a compact map.

This phenomenon is a "nonlinear analogue" of the invariance of the index of Fredholm operators by perturbation.

To describe the invariant, let us recall the (co)-homology of our infinite dimensional manifold $\mathcal{B}(M,P)$. For simplicity, we work over $\mathbf{Q}$-coefficient and assume that $\pi_1(M) = 1$. Let $b_2(M) = b_2 = \mathrm{rank}\, H_2(M;\mathbf{Q})$, and $[\Sigma_i]$, $i = 1, \cdots, b_2$ be its basis. (Here Σ_i are surfaces in M.)

Theorem 1.3 : (Donaldson [D3,D4])
(1)   *There exists a homomorphism* $\mu : H_2(M;\mathbf{Z}) \to H^2(\mathcal{B}(M,P);\mathbf{Z})$, *which induces an isomorphism over* $\mathbf{Q}$-*coefficient.*
(2)   *We put* $x_i = \mu(\Sigma_i)$. *Then* $H^2(\mathcal{B}(M,P);\mathbf{Q})$ *is isomorphic to the polynomial ring* $\mathbf{Q}[x_1, \cdots, x_{b_2}, y]$. *Here* y *is another generator of degree 4.*

We will discuss the construction of μ later. Now we put

"Definition" 1.4

$$Q_\ell(\Sigma_1,\cdots,\Sigma_\ell) = [\mathcal{M}(M,P)] \cap (x_1 \cup \cdots \cup x_\ell) \in \mathbf{Z}.$$

Here $2\ell = \dim \mathcal{M}(M,P)$.

To regard Donaldson invariant as a topological field theory, we need to study the case when our 4-manifold has a boundary N. Let $\mathcal{M}(M,P)$ be again the quotient of $\mathcal{A}(M,P)$ (= the set of all $SU(2)$ connections of $P \to M$) by $\mathcal{G}(M,P)$ (= the set of all gauge transformations.) Note that we assumed no boundary conditions here. Define $\mathcal{A}(N,P)$, $\mathcal{G}(N,P)$, $\mathcal{B}(N,P)$ in the same way. By restriction of connections, we obtain a map

$$res : \mathcal{B}(M,P) \to \mathcal{B}(N,P).$$

We put

$$\mathcal{M}(M,P) = \left\{ [A] \in \mathcal{B}(M,P) \,\middle|\, *F_A = -F_A \right\}.$$

Here we remark again that no boundary condition is assumed. As a consequence, the moduli space $\mathcal{M}(M,P)$ is infinite dimensional. In fact, we may regard :

$$\dim \mathcal{M}(M,P) - \frac{1}{2} \cdot \dim \mathcal{B}(M,P) \text{ is finite.}$$

We "put" $\infty = \dim \mathcal{B}(M,P)$. Then we :

"Definition" 1.5. Relative Donaldson invariant of M is $res_* [\mathcal{M}(M,P)] \in H_{\infty/2+finite}(\mathcal{B}(M,P))$.

Of course, the expression $H_{\infty/2+finite}$ does not make sense. But we can justify it by using Floer homology. (See §§ 5,6.)

We want to make this "definition" a bit more precise. Let $\mu : H_2(M;\mathbf{Q}) \to H^2(\mathcal{B}(M,P);\mathbf{Q})$ be as before. (This map is defined in the case when M has a boundary also.) For $\Sigma_i \subseteq M$ (surface), we choose $X_i \subset \mathcal{B}(M,P)$ codimension 2 submanifold (in the infinite dimensional manifold $\mathcal{B}(M,P)$) which represents the Poincaré dual to $\mu(\Sigma_i)$. Then we put

"Definition" 1.6.

$$Q_\ell(\Sigma_1, \cdots, \Sigma_\ell) = res_*([\mathcal{M}(M,P)] \cap (X_1 \cap \cdots \cap X_\ell)) \in H_{\infty/2 + finite}(\mathcal{B}(M,P)).$$

Finally we discuss the coupling formula. We consider the following situation. Let M_1, M_2 be 4-dimensional manifolds such that $\partial M_1 = -\partial M_2 = N$. (Here $-\partial M_2$ is ∂M_2 equipped with the opposite orientation.) We put $M = \partial M_1 \cup_N \partial M_2$. We assume furthermore $H_1(N; \mathbf{Z}) = 0$. This implies that $H_2(M; \mathbf{Z}) = H_2(M_1; \mathbf{Z}) \oplus H_2(M_2; \mathbf{Z})$. Let

$$\begin{cases} \Sigma_i \subseteq M_1, i = 1, \cdots, \ell \\ \Sigma_i' \subseteq M_2, i = 1, \cdots, \ell' \end{cases}$$

We want to find

$$Q_{\ell+\ell'}(M; \Sigma_1, \cdots, \Sigma_\ell, \Sigma_1', \cdots, \Sigma_{\ell'}')$$

in terms of

$$Q_\ell(M_1; \Sigma_1, \cdots, \Sigma_\ell)$$

and

$$Q_{\ell'}(M_2; \Sigma_1', \cdots, \Sigma_{\ell'}').$$

We remark that we may regard

$$\mathcal{B}(M,P) = \left\{([A],[B]) \in \mathcal{B}(M_1,P) \times \mathcal{B}(M_2,P) \,\middle|\, res([A]) = res([B])\right\}.$$

In other words

$$\mathcal{B}(M,P) = \mathcal{B}(M_1,P) \times_{\mathcal{B}(N,P)} \mathcal{B}(M_2,P).$$

Similarly we have

$$\mathcal{M}(M,P) = \mathcal{M}(M_1,P) \times_{\mathcal{M}(N,P)} \mathcal{M}(M_2,P).$$

Remark.   By the formula above we see that

$$\dim \mathcal{M}(M,P) = \dim \mathcal{M}(M_1,P) + \dim \mathcal{M}(M_2,P) - \dim \mathcal{M}(N,P).$$

is a finite number. This is consistent with our assertion that

$$\dim \mathcal{M}(M,P) - \frac{1}{2} \cdot \dim \mathcal{B}(N,P) \text{ is finite.}$$

Let

$$\mu_i : H_2(M_i; \boldsymbol{Q}) \to H^2(\mathcal{B}(M_i,P); \boldsymbol{Q}), \qquad i = 1,2,$$

be as before. We choose

$$X_j \subset \mathcal{B}(M_i,P), \quad \begin{cases} i = 1 & \text{if } j \le \ell, \\ i = 2 & \text{otherwise} \end{cases}$$

which represents the Poincaré dual to $\mu_1(\Sigma_j) \in H^2(\mathcal{B}(M_1,P); \boldsymbol{Q})$, $j \le \ell$, and $\mu_2(\Sigma'_{j-\ell}) \in H^2(\mathcal{B}(M_2,P); \boldsymbol{Q})$, $j > \ell$. Then $X_j \times_{\mathcal{B}(N,P)} \mathcal{B}(M_2,P) \subset \mathcal{B}(M,P)$ (or $\mathcal{B}(M_1,P) \times_{\mathcal{B}(N,P)} X_j \subset \mathcal{B}(M,P)$) represents the Poincaré dual to $\mu(\Sigma_j) \in H^2(\mathcal{B}(M,P); \boldsymbol{Q})$ or $\mu(\Sigma'_{j-\ell}) \in H^2(\mathcal{B}(M,P); \boldsymbol{Q})$.

Thus we have

$$
\begin{aligned}
Q_{\ell+\ell'}&(M; \Sigma_1, \cdots, \Sigma_\ell, \Sigma'_1, \cdots, \Sigma'_{\ell'}) \\
&= [\mathcal{M}(M,P)] \cap \bigcap_{j=1}^{\ell} [X_j \times_{\mathcal{B}(N,P)} \mathcal{B}(M_2,P)] \cap \bigcap_{j=\ell+1}^{\ell+\ell'} [\mathcal{B}(M_1,P) \times_{\mathcal{B}(N,P)} X_j] \\
&= \left[\mathcal{M}(M_1,P) \times_{\mathcal{B}(N,P)} \mathcal{M}(M_2,P) \right] \cap \bigcap_{j=1}^{\ell} [X_j \times_{\mathcal{B}(N,P)} \mathcal{B}(M_2,P)] \cap \bigcap_{j=\ell+1}^{\ell+\ell'} [\mathcal{B}(M_1,P) \times_{\mathcal{B}(N,P)} X_j] \\
&= res\left[\mathcal{M}(M_1,P) \cap \bigcap_{j=1}^{\ell} [X_j] \right] \bullet res\left[\mathcal{M}(M_2,P) \cap \bigcap_{j=\ell+1}^{\ell+\ell'} [X_j] \right] \\
&= Q_\ell(M_1; \Sigma_1, \cdots, \Sigma_\ell) \bullet Q_{\ell'}(M_2; \Sigma'_1, \cdots, \Sigma'_{\ell'}).
\end{aligned}
$$

Here $\bullet$: $H_{\infty/2+k}(\mathcal{B}(N,P)) \times H_{\infty/2-k}(\mathcal{B}(N,P)) \to \boldsymbol{Z}$ is the intersection product.

Of course, the calculations above are based on "intersection theory of cycles of infinite dimension", which is not established yet.

We thus described a heuristic argument to show the following coupling formula :

$$Q_{t+r'}(M;\Sigma_1,\cdots,\Sigma_{t+r'}) = Q_t(M_1;\Sigma_1,\cdots,\Sigma_t) \bullet Q_{r'}(M_2;\Sigma_t,\cdots,\Sigma_{t+r'}).$$

This formula plays an important role in topological field theory. See Atiyah [A] for more discussion on topological field theory.

Thus we described very rough ideas to define Donaldson invariant and to regard it as a topological field theory in dimension 3 + 4. It has a lot of applications to 4-dimensional topology. Namely, using Donaldson invariant one can find a lot of pairs of 4-manifold homeomorphic but not diffeomorphic to each other. But here, because mainly of the ability of the author, we do not discuss it. Instead we discuss more about the general results on the solution of ASD equation and how they are used to realize the ideas illustrated in this section.

§ 2 Basic properties of the moduli space of ASD connections

In this section, we want to make some part of the discussions of the previous section slightly more rigorous, and want to explain some results required to justify the constructions. There are basically three points to be clarified.

(A)   **Compactness (or compactification) of the moduli spaces**
           $\mathcal{M}(M,P)$,

(B)   **Singularity of $\mathcal{B}(M,P)$,**

(C)   **Dimension of $\mathcal{M}(M,P)$.**

<u>(C)   Index theorem</u>

Let us first discuss the third point. The rough counting of freedom in the last section shows that $\mathcal{M}(M,P)$ is finite dimensional (in the case when M is compact). This rough counting can be justified by using the theory of elliptic complex or Atiyah-Singer index theorem. This is a linear part of the story. That is, to count the dimension we are only to study the linearization of the nonlinear equation $F_A^+ = 0$. Let us do it here. We calculate

$$
\begin{aligned}
F_A &= dA + A \wedge A, \\
F_{A+\delta A} - F_A &= d(\delta A) + A \wedge \delta A + \delta A \wedge A + O(|\delta A|^2) \\
&= d_A(\delta A) + O(|\delta A|^2).
\end{aligned}
$$

Hence putting $\tilde{\mathcal{M}}(M,P) = \left\{ A \mid F_A^+ = 0 \right\}$, we get

$$
T_A \tilde{\mathcal{M}}(M,P) = \left\{ B \in \Gamma(\Lambda^1 \otimes su(2)) \mid d_A B + *d_A B = 0 \right\}.
$$

We put $P_+ d_A = \dfrac{1}{2}(d_A + *d_A)$. Then

(2.1) $T_A \tilde{\mathcal{M}}(M,P) = \operatorname{Ker} P^+ d_A.$

Next we recall

$$
T_A \mathcal{B}(M,P) = T_A \big(\mathcal{G}(M,P) \cdot A \big)^\perp.
$$

Here $\perp$ denotes the orthonormal complement and $\mathcal{G}(M,P) \cdot A$ the gauge orbit of A. We recall that the Lie algebra of $\mathcal{G}(M,P)$ is identified with

$$\Gamma(M, \Lambda^0 \otimes Ad\ P) = \{su(2) \text{ - valued 0 - forms, twisted by } P\}.$$

Let $u \subset \Gamma(M, \Lambda^0 \otimes \Lambda d\ P)$. Then

$$\frac{d}{dt}\left((\exp tu)^* A\right)\Big|_{t=0} \quad = \frac{d}{dt}(1 - tu) \circ (d + A) \circ (1 + tu)\Big|_{t=0}$$
$$= du + Au - Au = d_A u.$$

Thus

$$T_A \mathcal{B}(M, P) = \operatorname{coker} d_A = \operatorname{Ker} d_A^* \subset \Gamma(M, \Lambda^1 \otimes AdP).$$

We obtained the following complex :

$$0 \to \Lambda^1 \otimes AdP \xrightarrow{P_+ d_A \oplus d_A^*} (\Lambda_+^2 + \Lambda^0) \otimes AdP \to 0$$

This complex is called Atiyah-Hitchin-Singer complex ([S], [AHS]) . The dimension of our moduli space $\mathcal{M}(M, P)$ is equal to that of the kernel of this complex.

In general, the cokernel of this complex does not vanish. But, by a fairly standard transversality argument, we can perturb our section $[A] \mapsto [F_A^+]$ such that the corresponding complex is surjective, provided our space $\mathcal{M}(M, P)$ is nonsingular at $[A]$. (We will discuss this point later.) Moreover, the remark just before Theorem 1.3 shows that the section $[A] \mapsto [F_A^+]$ can be perturbed without changing the resulting invariant. Hence, here we simply forget this point and suppose

$$\dim \mathcal{M}(M, P) = \text{Index of Atiyah - Hitchin - Singer complex.}$$

We can then calculate the index and obtain :

Theorem 2.2 ([S], [AHS]).
$$\dim \mathcal{M}(M,P) = 8k - 3(1 + b_+^2 - b_1).$$

Here $k = c^2(P) \cap [M]$, $b_1 = \operatorname{rank} H_1(M;\boldsymbol{Q})$, *and* b_2^+ *is the number of positive eigenvalues of the intersection pairing :* $H_2(M;\boldsymbol{Q}) \otimes H_2(M;\boldsymbol{Q}) \rightarrow \boldsymbol{Q}$.

We do not try to explain the calculation of the index here. Instead, let me mention a way to remember this formula. First consider the case when P is trivial. Then

$$\begin{cases} \operatorname{Ker}(P_+ d \oplus d^*) = H^1(M;\boldsymbol{R}) \otimes su(2) \\ \operatorname{coker} d^* = H^0(M;\boldsymbol{R}) \otimes su(2) \\ \operatorname{coker} P_+ d = \left\{ [u] \in H^2(M;\boldsymbol{R}) \big| * u = u \right\} \otimes su(2) \end{cases}$$

Hence the index of Atiyah-Hitchin-Singer complex is $3(b_1 - 1 - b_2^+)$. Thus we obtained the formula in the case $k = 0$. So all one needs to remember is the number 8, the coefficient of $k = c^2(P) \cap [M]$.

(B) Singularity of $\mathcal{B}(M,P)$

We next turn to the second point. We mention first why a singularity can cause a trouble for the construction in the last section.

We recall that the construction explained in § 1 is basically to consider intersection numbers of various cycles in (infinite dimensional) "manifold" $\mathcal{B}(M,P)$.   (Namely to calculate $[\mathcal{M}(M,P)] \cap \bigcap_{i=1}^{\ell} [X_i]$, where $[X_i]$ is the Poincaré dual to $\mu(\Sigma_i)$.) Now we explain, by a simple example in finite dimensional case, that intersection theory has serious trouble on a singular space.

First, we put $X = \boldsymbol{R}^7$ (a smooth manifold). And let $Z_1, Z_2 \subset X$ be 3-dimensional submanifolds. Then $3 + 3 - 7 = -1 < 0$ implies that we can perturb them so that $Z_1 \cap Z_2 = \varnothing$.

Next, we put $X = $ cone of $\boldsymbol{CP}^3$, a singular space of dimension 7. Let $S_i^2 \subset \boldsymbol{CP}^3$, $i = 1,2$, be spheres representing nontrivial cohomology classes. Let Z_i $i = 1,2$, be the cone of S_i^2. Then $Z_i \subset X$ are cycles of dimension 3, which intersect to each other at one point. We can show easily that we can *not* perturb Z_i $i = 1,2$ such that $Z_1 \cap Z_2 = \varnothing$, though

"virtual dimension" of $Z_1 \cap Z_2 = 3 + 3 - 7 = -1 < 0$.

In the discussion of dimension of $\mathcal{M}(M,P)$, we mentioned that we can perform perturbation so that $\dim \mathcal{M}(M,P) = $ index of Atiyah Hitchin Singer complex (= virtual dimension) *provided the point in question is smooth* By the same reason as the above example, this is no longer true at a singular point. Namely, the space of negative virtual dimension may not be empty. This causes various troubles, for example, to the well-definedness of the intersection number. (That is, to prove its independence of the perturbation.)

Now we discuss the singularities of the space $\mathcal{B}(M,P)$, which is the quotient of the (infinite dimensional) affine space $\mathcal{A}(M,P)$ by the action of gauge transformation group. Hence its singularities coincide with the set of the fixed points of the action. To study the fixed point, let us consider the space $\tilde{\mathcal{B}}(M,P)$. We fix a point $x_0 \in M$ and put

$$G_0(M,P) = \left\{ g \in \mathcal{G}(M,P) \,\middle|\, g(x_0) = id \right\}.$$

We remark that the action of $G_0(M,P)$ on $\mathcal{A}(M,P)$ is free, hence the quotient space $\tilde{\mathcal{B}}(M,P) = \mathcal{A}(M,P)/G_0(M,P)$ is a smooth manifold (of infinite dimension.) On $\tilde{\mathcal{B}}(M,P)$ the group $SU(2) = \mathcal{G}(M,P)/G_0(M,P)$ acts such that $\mathcal{B}(M,P) = \tilde{\mathcal{B}}(M,P)/SU(2)$. We remark that the action of $\{\pm 1\} \subset SU(2)$ is trivial. A singularity of $\mathcal{B}(M,P)$ coincides with the point where the isotropy group of the action is strictly larger than $\{\pm 1\}$. For $[A] \in \tilde{\mathcal{B}}(M,P)$ we put

$$I_A = \left\{ g \in \mathcal{G}(M,P) \,\middle|\, g[A] = [A] \right\}.$$

We can describe this group as follows. Let A be a connection and $\ell : (S^1, 1) \to (M, x_0)$ be a loop. Then we obtain a holonomy $h_\ell(A) : P_{x_0} \to P_{x_0}$. We may identify $h_\ell(A) \in (\mathcal{G}(M,P)/G_0(M,P)) = SU(2)$. Then we define the holonomy group $\mathrm{Hol}(A) \subseteq SO(3)$ to be the set of all holonomy maps $\mathrm{Hol}(A) \subseteq SU(2)$. Now we find that

$$I_A \cong \text{Centralizer of } \mathrm{Hol}(A) = \left\{ g \in SU(2) \,\middle|\, gh_\ell(A) = h_\ell(A)g, \text{ for all } \ell \right\}.$$

We say that A is a reducible connection if I_A is larger than $\{\pm 1\}$. Now there are two kinds of reducible connections.

(1) $I_A = SU(2)$.

(2) $I_A = U(1)$.

(1) is the case when $\mathrm{Hol}(A) \subseteq \{\pm 1\}$. When we assume that $H_1(M;\mathbf{Z}) = 0$, there is only one such a connection, that is the trivial connection.

(2) is the case when $\mathrm{Hol}(A) \not\subset U(1)$. Let us discuss this case. The group $U(1)$ is embedded into $SU(2)$ as

$$\left\{ \begin{pmatrix} \lambda & 0 \\ 0 & \lambda^{-1} \end{pmatrix} \middle| \lambda \in \mathbf{C}, |\lambda| = 1 \right\}.$$

Then we put

$$L_x = \left\{ \mathrm{Par}_\ell \begin{pmatrix} a \\ 0 \end{pmatrix} \middle| \begin{matrix} a \in \mathbf{C}, \\ \ell : [0,1] \to M, \ell(0) = x_0, \ell(1) = x \end{matrix} \right\} \subseteq (P \times_{SU(2)} \mathbf{C}^2)_x.$$

Here Par_ℓ denotes the parallel translation along ℓ by our connection A. L is a complex line bundle and $P \times_{SU(2)} \mathbf{C}^2 \cong L \oplus L^*$ as $U(1)$-bundle. The connection A on P also splits into a connection on L and its dual connection on L^*. Namely

$$A = \begin{pmatrix} A_0 & 0 \\ 0 & -A_0 \end{pmatrix}.$$

Then

$$F_A = \begin{pmatrix} dA_0 & 0 \\ 0 & -dA_0 \end{pmatrix}$$

(We remark here that A_0 is not a global 1 form (unless L is a trivial bundle), but dA_0 is a global 2-form.)

Now we have

$$*F_A = -*F_A \quad \Leftrightarrow \quad *dA_0 = -dA_0 \quad \Leftrightarrow \quad d^*dA_0 = 0.$$

We have also $ddA_0 = 0$. Hence, A is self-dual if and only if dA_0 is a harmonic 2 form.

On the other hand, we have

$$c^1(L) = \frac{1}{2\pi}[dA_0],$$

$$c^2(P) - \frac{1}{4\pi^2}\det F_A.$$

Hence

$$c^2(P) = \frac{-1}{4\pi^2}dA_0 \wedge dA_0.$$

Furthermore, remark that $u = dA_0$ for some connection A_0 of L if and only if $du = 0$ and $[u] = c^1(L) \in H^1(M;\mathbf{Z})$.

Thus we have

Lemma 2.4.

$$\left\{ A \in \mathcal{A}(M,P) \,\middle|\, \begin{array}{l} I_A = U(1) \\ F_A = -*F_A \end{array} \right\} \middle/ \mathcal{G}(M,P) = \left\{ u \in \Lambda^1(M) \,\middle|\, \begin{array}{l} du = 0, \\ *u = -u, \\ [u \wedge u] = c^2(P), \\ [u] \in H^2(M;\mathbf{Z}). \end{array} \right\}.$$

By this lemma we can describe the ASD connection with $I_A = U(1)$ in terms of the cohomology and harmonic theory of M.

(A) Compactness of the moduli space

The following theorem (due to Uhlenbeck) plays a key role in compactification of the moduli space of ASD connections.

Theorem 2.5 (Uhlenbeck [U1]). *Let* A *be an* $SU(2)$ *connection on* $D^4 - \{0\}$. *Suppose*

(2.5.1) $*F_A = -F_A$,

(2.5.2) $\int \|F_A\|^2 < \infty$.

Then there exists a gauge transformation $g : D^4 - \{0\} \to SU(2)$ *such that* g^*A *extends to a smooth connection on* D^4.

We do not discuss the proof of this theorem here. (However we give a few comments about it in § 6.)

We will apply it to show the following :

Corollary 2.6.   *Let A be a connection on $\mathbf{R}^4$ such that*

(2.6.1) $*F_A = -F_A,$
(2.6.2) $\int \|F_A\|^2 < \infty.$

Then

$$\int \|F_A\|^2 \geq 4\pi^2.$$

For the proof we first remark that the ASD equation (2.6.1) is conformal invariant. Namely, if A satisfies (2.6.1) for a metric g, then it also satisfies (2.6.1) for the metric fg for any positive function f.

We next remark that $\mathbf{R}^4$ is conformally equivalent to $S^4 - \{\text{one point}\}$. Hence by Theorem 2.5, there is an $SU(2)$-bundle P, a connection $\hat{A}$ on it, and a bundle isomorphism $\varphi : P|_{\mathbf{R}^4} \cong \mathbf{R}^4 \times SU(2)$ such that $\varphi^*(A) = \hat{A}|_{\mathbf{R}^4}$. Hence

$$\int_{\mathbf{R}^4} \|F_A\|^2 = \int_{S^4} \|F_{\hat{A}}\|^2 = 4\pi^2 [c^2(P)] \cap [S^4] \in 4\pi^2 \mathbf{Z}.$$

The corollary follows.

The most important result on the compactness of the moduli space of ASD connections is the following :

Theorem 2.7 ([U1]).   *Let $P \to M$ be an $SU(2)$-bundle and A_i be a sequence of ASD connections on P. Then there exists a subsequence A_{i_j} such that $\lim\limits_{j \to \infty} |F_{A_{i_j}}|$ diverges at most finitely many points on M.*

Theorem 2.7 bis.   *More precisely, there exist $\{p_1, \cdots, p_N\} \subset M$, a connection A_∞ on $M - \{p_1, \cdots, p_N\}$ and $n_i \in \mathbf{Z}_+$ such that*

$$\lim_{j \to \infty} |F_{A_{i_j}}| = |F_{A_\infty}| + \sum 4\pi^2 n_i \delta_{p_i}$$

as measure and that A_∞ is extended to a connection on $P' \to M$ such that

$$c^2(P) = c^2(P') + \sum n_i.$$

Sketch of the proof : We discuss only the proof of Theorem 2.7. Suppose $\lim_{i \to \infty} |F_{A_i}|(p_k) = \infty$ at $p_1, \cdots, p_J$. It suffices to estimate J. We put $C_{i,k} = |F_{A_i}|(p_k)$. Define a map $\varphi_{i,k} : B^4(C_{i,k}) \to M$ by $\varphi_{i,k}(x) = \exp_{p_k}(x / \sqrt{C_{i,k}})$. Then $|\varphi_{i,k}^*(A)(o)| = 1$. We may assume that, after taking a subsequence, $\lim_{i \to \infty} \varphi_{i,k}^*(A)$ converges to a nontrivial ASD connection on $\mathbf{R}^4$. Hence Corollary 2.6, implies that

$$\lim_{i \to \infty} \int_{B(p_k, 1/\sqrt{C_{i,k}})} \|A\|^2 = \lim_{i \to \infty} \int \left\| \varphi_{i,k}^*(A) \right\|^2 \geq 4\pi^2.$$

Therefore

$$4\pi^2 c^2(P) = \limsup_{i \to \infty} \int_M \|A_i\|^2 \geq \sum_{k=1}^J \lim_{i \to \infty} \int_{B(p_k, 1/\sqrt{C_{i,k}})} \|A_i\|^2 = \sum_{k=1}^J \lim_{i \to \infty} \int \left\| \varphi_{i,k}^*(A) \right\|^2 \geq 4\pi^2 J.$$

It follows that $c^2(P) \geq J$ as required.

Construction of Donaldson invariant

Now we use the above results on the moduli space of ASD connections to make the definition of Donaldson invariant in the last section a bit more rigorous.

First we recall the definition of the map $\mu : H_2(M^4 ; \mathbf{Z}) \to H^2(\mathcal{B}(M, P) ; \mathbf{Z})$. Let $x \in H_2(M^4 ; \mathbf{Z})$ and Σ^2 be the surface representing x. We define a complex line bundle $\mathcal{L}(\Sigma)$ on $\mathcal{B}(\Sigma, P)$ by

$$\mathcal{L}(\Sigma)_{[A]} = \Lambda^{top} \left(\mathrm{Ker}(\overline{\partial}_A : \Lambda^{0,0}(\Sigma) \otimes P \to \Lambda^{0,1}(\Sigma) \otimes P) \right)^* \otimes$$
$$\Lambda^{top} \left(\mathrm{CoKer}(\overline{\partial}_A : \Lambda^{0,0}(\Sigma) \otimes P \to \Lambda^{0,1}(\Sigma) \otimes P) \right).$$

Here $\overline{\partial}_A$ is the Dolbeault operator twisted by the connection A. And for a vector space E, the space $\Lambda^{top} E$ denotes $\Lambda^{\dim E} E$.

Let $res_\Sigma : \mathcal{B}(M, P) \to \mathcal{B}(\Sigma, P)$ be the restriction map. Then by definition $\mu(x) = res_\Sigma^*(c^1(\mathcal{L}(\Sigma)))$. More explicitly, we choose a generic section s_Σ to $\mathcal{L}(\Sigma)$, and the

Poincaré dual to $\mu(x)$ is $res_\Sigma^{-1}(s_\Sigma^{-1}(0))$.

Now we suppose

$$2\ell = \dim \mathcal{M}(M,P) = 8c^2(P) - 3(1 + b_2 - b_1).$$

Let $x_i \in H_2(M;\mathbf{Z})$, $i = 1,\cdots,\ell$, and $x_i = [\Sigma_i]$. Then the definition in the last section can be written as

$$(2.8) \qquad Q_\ell(x_1,\cdots,x_\ell) = [\mathcal{M}(M,P)] \cap \bigcap_{i=1}^{k} [res_{\Sigma_i}^{-1}(s_{\Sigma_i}^{-1}(0))].$$

Now, the first trouble to justify Formula (2.8) is that $\mathcal{M}(M,P)$ is noncompact and hence its fundamental class is not well-defined. This difficulty is overcome by using Theorem 2.7 as follows.

Let $S^k(M) = M^k / S_k$, where S_k is the symmetric group of order $k!$ acting on M^k as the permutation of components. Let $c^2(P) = m$ and let P_n denote the $SU(2)$ bundle with $c^2(P_n) = n$. We put

$$C\mathcal{M}(M,P) = \mathcal{M}(M,P) \cup \bigcup_{n=1}^{m} S^n(M) \times \mathcal{M}(M,P_{m-n}).$$

We define a topology on $C\mathcal{M}(M,P)$ as follows. We identify $S^n(M) = \left\{ \sum_{i=1}^{m} n_i \delta_{p_i} \mid \Sigma n_i = n \right\}$. Then we put

$$\lim_{n \to \infty} A_i = (\Sigma n_i \delta_{p_i}, A)$$

if

$$\lim_{n \to \infty} \|F_{A_i}\| = \frac{1}{4\pi^2} \Sigma n_i \delta_{p_i} + \|F_A\|,$$

$$\lim_{n \to \infty} A_i = A \qquad \text{on } M - \{p_1,\cdots,p_m\}.$$

Then, by Theorem 2.7, the closure $\overline{\mathcal{M}(M,P)}$ of $\mathcal{M}(M,P)$ in $C\mathcal{M}(M,P)$ is compact.[1] On the other hand

---

[1] In the case when M is Kähler, J. Lie recentluy proved that $\overline{\mathcal{M}(M,P)}$ is a projective variety.

$$\dim S^n(M) \times \mathcal{M}(M, P_{m-n}) = 4n + \dim \mathcal{M}(M, P_{m-n})$$
$$= -4n + \dim \mathcal{M}(M, P_m),$$

Therefore

(2.9) $$\dim\left(\overline{\mathcal{M}(M,P)} - \mathcal{M}(M,P)\right) \leq \dim \mathcal{M}(M,P) - 2.$$

Hence the fundamental class $[\mathcal{M}(M,P)]$ is well-defined.

In fact, there are two cheatings in the above argument. First the class $\mu(x_i)$ on $\mathcal{B}(M,P)$ or on $\mathcal{M}(M,P)$ does not extend to $\overline{\mathcal{M}(M,P)}$ in general. Hence the above naive calculation of the dimension is not sufficient to prove the well-definedness of (2.8).

Second because of the presence of trivial connection the formula (2.9) may not hold in case $m - n = 0$.

We do not discuss these points here and only mention that we need to assume that $c^2(P)$ is sufficiently large to show the well definedness of the invariant in this way.

Now we come to the second problem, the singularity of the space $\mathcal{B}(M,P)$. Since the Chern class of P is nonzero, there is no trivial connection in $\mathcal{B}(M,P)$. But we need to consider the singular point for which $I_A = U(1)$. In (B) we found that the ASD connections on this singular locus is parametrized by

$$\Xi = \left\{ u \in \Lambda^2(M) \left| \begin{array}{l} du = 0, \\ *u = -u, \\ [u \wedge u] = c^2(P), \\ [u] \in H^2(M;\mathbf{Z}). \end{array} \right. \right\}.$$

So we want to discuss when this set is nonempty. We consider the space $\mathcal{H}^2(M;\mathbf{R}) = \left\{ u \in \Lambda^2(M) \,\middle|\, du = d^*u = 0. \right\}$ of harmonic 2 forms. We consider the following three cases separately.

(Case 1) $b_2^+ = 0$.

In this case, $\Xi = \mathcal{H}^2(M;\mathbf{R})$. Hence Ξ is nonempty unless $b_2 = 0$. Thus in this case there is always a singular point. So the way we have explained to define Donaldsons invariant breaks down.

(Case 2) $b_2^+ = 1$.

In this case, we can identify $\mathcal{H}^2(M;\mathbf{R})$ (equipped with intersection product) with the Laurentz space $\left(\mathbf{R}^{n+1}, x_0^2 - \sum_{i=1}^{n} x_i^2 \right)$. We put $\Psi = \left\{ u \in \mathcal{H}^2(M;\mathbf{R}) \mid [u \wedge u] = 1 \right\}$ this space is identified with the hyperboloid with two components. (See Figure 2.10.) Let Ψ_+ be one of the components.

The Hodge $*$ operator preserves $\mathcal{H}^2(M;\mathbf{R})$. Hence there exists x_0 with $x_0 \bullet x_0 > 0$ such that $*u = -u + 2(x_0 \bullet u)x_0$. There is a uniqu such element x_0 in Ψ_+.

Next for each $a \in H^2(M;\mathbf{Z})$, $a \cup -a = c^2(P)$, we put $\Pi_a = \left\{ x \in \Psi_+ \mid [x \wedge a] = 0 \right\}$. Π_a is a codimension one submanifold and $\bigcup_a \Pi_a$ devide Π_a into the union disjoint and locally finite family of domains.

Hence $\bigcup_a \Pi_a$ is a union of codimension one (totally geodesic) submanifold of the hyperboloid.

We remark that the Hodge $*$ operator and, hence, x_0 moves as we perturb Riemannian metrics.

Consider the set

$$\Xi = \left\{ u \in \Lambda^2(M) \;\middle|\; \begin{array}{l} du = 0, \\ *u = -u, \\ [u \wedge u] = c^2(P), \\ [u] \in H^2(M;\mathbf{Z}). \end{array} \right\}.$$

We remark also that Ξ is nonempty if and only if x_0 is contained in $\bigcup_a \Pi_a$. Therefore, in this case, x_0 the set of the metric such that Ξ is nonempty is codimension one subset of the set of all metrics.

So Donaldson invariant is well-defined for a generic metric. But there is a trouble to prove its independence of the metric. Namely, suppose we have two metrics g_1, g_2 such that $\Xi = \emptyset$. Then to show that Donaldson invariant obtained using these two metrics coincide, we need to join these two metrics by a path. But on this path there may be a metric such that $\Xi \neq \emptyset$. Thus in general Donaldson invariant *does* depend on the choice of the metric. But, in fact, we can describe the way how they depends, (using the hyperboloid picture as above.) The chamber structure discussed in Donaldson [D5] is related to this problem. See [K], for more discussion on this point.

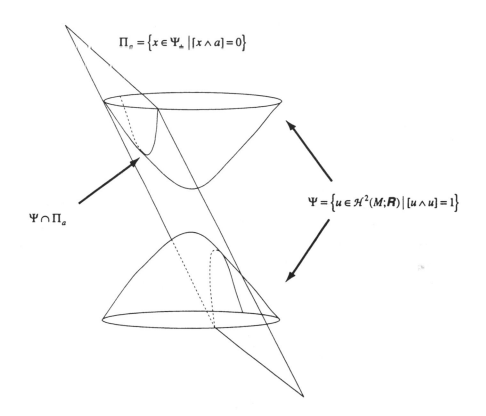

Figure 2.10

(Case 3) $b_2^+ \geq 2$.

By using the discussion above, we find that the trouble occurs on codimension $b_2^+ \geq 2$ subset of metrics. Hence in this case there is no trouble caused by the singularity with $I_a = U(1)$. (However if we use the moduli space of ASD connections to define a characteristic class of a fibre bundle with 4-manifold fibres, this will again cause a similar problem.)

§ 3 Casson invariant and gauge theory

Now we consider a 3-dimensional manifold N. We first discuss Casson's invariant. (See [AK] for the rigorous definition of Casson invariant.) We consider the fundamental group $\pi_1(N)$ of N and denote it by $R(M)$ the set of all representations of $\pi_1(N)$ in $SU(2)$ divided by conjugation. Then roughly speaking we have

$$\text{"Casson invariant of } N \quad = \quad \frac{\chi(R(N))}{2} \text{"}.$$

Typically the space $R(N)$ is 0-dimensional. Then most naively speaking its Euler number is its order. But it is natural to consider $R(N)$ as a 0-dimensional oriented manifold.[1] So Casson invariant is the difference between the number of the points of $R(N)$ with positive sign and one with negative sign.

We define an orientation of $R(N)$ as follows. Let $N = H_g^1 \cup_{\Sigma_g} H_g^2$ be the Heegaard splitting of N. Namely, H_g^i are the handle bodies , and we patch them along their boundaries Σ_g (the surface of genus g) to get the 3-manifold N. We consider the representation space $R(\Sigma_g)$ of the fundamental group of Σ_g. One can prove that $R(\Sigma_g)$ has a natural orientation. (For example a complex structure on Σ_g induces one on $R(\Sigma_g)$, hence an orientation on it.) Next we consider the map $I_i : R(H_g^i) \to R(\Sigma_g)$ induced by the inclusion. By van-Kampen's theorem we have

$$R(N) = I_1\!\left(R(H_g^1)\right) \cap I_2\!\left(R(H_g^2)\right).$$

Let us fix an orientation of $R(H_g^1)$. Then we remark that the orientability of N implies that there exists an orientation preserving diffeomorphism of $R(\Sigma_g)$ which sends the image $I_1\!\left(R(H_g^1)\right)$ to $I_2\!\left(R(H_g^2)\right)$. Hence the orientation of $R(H_g^1)$ induces one on $R(H_g^2)$. Thus we get the orientation on the intersection $R(N)$.

Thus we obtain an orientation on $R(N)$. By using an appropriate perturbation of embeddings I_i, we can define Casson invariant in the case when $R(N)$ is not necessarily discrete.

To show that it is an invariant of 3-manifold N one needs to show that the orientation discussed above is independent of the choice of the Heegard splitting. Casson proved it by finding explicitly how his invariant changes under the change of Heegard splitting. This approach has a big advantage that in this way one obtains an explicit algorithm to calculate

---

[1]Under assumptions which we do not specify here.

the invariant.   On the other, hand it has a disadvantage that it breaks natural symmetry by introducing Heegard splitting in rather artificial way [2]   Thus it is important to have aother approach which does not require artificial choices like Heegard splitting   Such an approach is provided by Taubes [T3] and is based on gauge theory.

As in § 1, let $\mathcal{A}(N)$ be the set of all $su(2)$-connections of the trivial $SU(2)$-bundle[3] on N, $\mathcal{G}(N)$ be the set of all gauge transformations, and $\mathcal{B}(N)$ the quotient of $\mathcal{A}(N)$ by $\mathcal{G}(N)$. Then, roughly speaking, we have

$$\text{"Casson invariant of } N \;=\; \frac{\chi(\mathcal{B}(N))}{2}."$$

The Euler number in the right hand side is not usual one.

Now, let us try to "prove" $\dfrac{\chi(R(N))}{2} \;=\; \dfrac{\chi(\mathcal{B}(N))}{2}$ and define it.

Let us recall the case of a finite dimensional manifold. Suppose that we have a vector field V on a manifold X. Let $p \in X$ with $V(p) = 0$. The covariant derivative $\nabla V(p)$ of V at p defines a linear map : $T_p X \to T_p X$. We say that p is nondegenerate critical point if $\nabla V(p)$ is invertible for all such p. For a nondegenerate critical point we define its index by

$$i_V(p) = (-1)^{\text{number of negative eigen values of } \nabla V(p)}.$$

We assume that all the critical points are nondegenerate. Then there are only a finite number of critical points. Then a classical result by Hopf says

$$\chi(X) = \sum_{p:\text{critical point}} i_V(p).$$

Now we apply this idea to our infinite dimensional situation. First we remark that we can identify

$$T_{[a]}\mathcal{B}(N) = \left\{ u \in \Gamma(\Lambda^1(N) \otimes su(2)) \,\middle|\, d_a^* u = 0 \right\}.$$

---

[2]The same phenomenon happened in the study of Jones-Witten invariant where the "definition" based on Feymann path integral and gauge theory is natural and show the basic reason for the invariance, while the approach using Dehn surgery, Heegard splitting, triangle decomposition etc. and representation theory gives an algorighm of calculation but breaks the natural symmetry of the theory.

[3]We remark that all $SU(2)$ bundles over N is trivial.

And by Bianci identity

$$d_a^*(*F_a) = -*d_a**F_a = \pm d_a F_a = 0.$$

Thus the map $a \mapsto *F_a$ defines a vector field on $\mathcal{B}(N)$. The set of critical points of this vector field is $R(N)$.

For a critical point $[a]$ of this vector field $*F_a$, we can calculate $\nabla *F_a$ as follows :

$$\nabla_u *F_a = \frac{d}{dt}*F_{a+tu}\bigg|_{t=0} = *d_a u.$$

Here we recall $u \in T_{[a]}\mathcal{B}(N) = \left\{ u \in \Gamma(\Lambda^1(N) \otimes su(2)) \,\middle|\, d_a^* u = 0 \right\}$. (Then automatically $*d_a u \in T_{[a]}\mathcal{B}(N) = \left\{ \dot{u} \in \Gamma(\Lambda^1(N) \otimes su(2)) \,\middle|\, d_a^* u = 0 \right\}$).

We assume that this operator is invertible for every flat connection a. (This is not, in fact, necessary, since it is always satisfied after a suitable perturbation of our function cs.)

Now the trouble is to define the index of a critical point in this infinite dimensional situation. In fact, one can prove that the number of negative eigenvalues of this operator $*d_a$ is infinite.

The way to overcome this difficulty is as follows. Let a, b be two flat connections. We can "define" the difference between the number of negative eigenvalues of $*d_a$ and one of $*d_b$ as follows.

Choose a path a_t in $\mathcal{A}(N)$ such that $a_0 = a$ and $a_1 = b$. We draw the eigenvalues of $*d_{a_t}$ as follows :

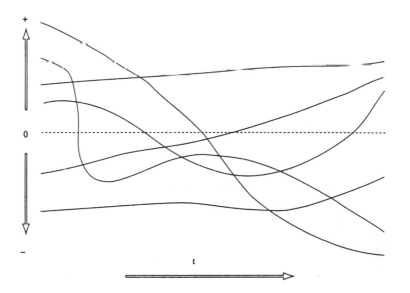

Figure 3.3

The lines above denote the eigenvalues at $t \in [0,1]$. (In fact, there are infinitely many lines.) The operator $*d_{a_t}$ has zero eigenvalue when a real line meets the dotted line. (This figure is called the spectral flow.) Now let us count the number of lines which cross dotted the line from $+$ to $-$ and from $-$ to $+$. (In the above example they are 3 and 2.) We *define*

$$i_{*F_a}(a) = i_{*F_a}(b) \qquad \text{if the difference of this number is even}$$
$$i_{*F_a}(a) \neq i_{*F_a}(b) \qquad \text{if the difference of this number is odd.}$$

(We recall that $i_{*F_a}(a) = \pm 1$.) In the above example, $i_{*F_a}(a) \neq i_{*F_a}(b)$, since $3-2$ is odd. One can prove that the above definition is independent of the choice of the path.

Hence, by fixing a sign at one flat connection, we can "define" the index $i_{*F_a}(a)$.

In this way one can "define" the Euler number $\dfrac{\chi(\mathcal{B}(N))}{2}$.[4] Taubes also proved that this number coincides with Casson invariant.

Let us remark that what we did here is quite similar to one in § 1. Here we use the zeros of a section of the tangent bundle $T\mathcal{B}(N)$ to define Casson's invariant, there we used the zeros of a section of the vector bundle $\mathcal{A}(M,P) \times_{\mathcal{G}(M,P)} \Gamma(\Lambda_+^2 \otimes \mathrm{Ad}P) \to \mathcal{B}(M,P)$ to define Donaldson's invariant.

---

[4] In fact, to do it, we need to assume that N is a homology sphere.

§ 4 Floer homology

In the previous section we utilized a vector field $*F_a$ on $\mathcal{B}(N)$ to "define" its Euler number. In our situation we can further prove that $*F_a$ is a gradient vector field. We will explain later. In general, when one has a vector field on a manifold, one can use it to calculate the Euler number of the manifold. But in the case when the vector field is a gradient vector field, one can do more and in fact can calculate the homology group.

For example, let us recall the following fact. Let X be a manifold of finite dimension and V be a vector field on it. Assume that all the singular points (the point where V vanishes) are nondegenerate. (Namely, the differential ∇V is invertible there.) Then a theorem by Hopf implies :

$$\#\left\{p \in X \mid V(p) = 0\right\} \geq |\chi(X)| = \left|\sum_{k=0}^{\dim X}(-1)^k \operatorname{rank} H^k(X;\boldsymbol{Q})\right|.$$

On the other hand, if V is a gradient vector field, we have the following stronger inequality called the Morse inequality :

$$\#\left\{p \in X \mid V(p) = 0\right\} \geq \sum_{k=0}^{\dim X} \operatorname{rank} H^k(X;\boldsymbol{Q}).$$

Thus the Morse index corresponds to the lift to $\boldsymbol{Z}$ of the index of the vector field $\in \boldsymbol{Z}_2 = \{\pm 1\}$.

In our case, our vector field $*F_a$ is a gradient vector field. Hence for $p \in R(N)$ its contribution $(\in \boldsymbol{Z}_2 = \{\pm 1\})$ to the index of vector field can be lifted to a Morse index $\in \boldsymbol{Z}$. (More precisely in the present situation our index is in $\boldsymbol{Z}_8$. This phenomenon has no finite dimensional analogy and will be explained a bit more later on.)

This observation on the lifting of index to $\boldsymbol{Z}$ is the linear part of the story. There is also a nonlinear part of the story which we are going to describe in this section.

Before doing it, let us discuss another point of view for the lifting of index. Let us go back to the discussion at the beginning of § 3. There we introduced the Heegard splitting $N = H_g^1 \cup_{\Sigma_g} H_g^2$ and remarked $R(N) = I_1\left(R(H_g^1)\right) \cap I_2\left(R(H_g^2)\right)$. For $p \in R(N)$, its contribution to the Casson invariant coincides to its contribution to the intersection number $I_1\left(R(H_g^1)\right) \bullet I_2\left(R(H_g^2)\right)$. Therefore it relates to the orientation or the first Stiefel-Whiteney class.

However, in our situation the submanifolds $I_1\left(R(H_g^i)\right)$, $i = 1, 2$, have further structure than orientation. Namely, they are Lagrangian submanifolds of the symplectic manifold

$R(N)$. In consequence, we can lift our Stiefel-Whiteney class to a torsion free class, the universal Maslov class. To see this relation more explicitly, we consider the following Grassmannians.

$$Gr_{2n,n} = \left\{ E \subset \mathbf{C}^n \mid n \text{ dimensional } \mathbf{R}\text{ - linear subspaces} \right\}$$

$$Lag_n = \left\{ E \in Gr_{2n,n} \mid \sum dx_i \wedge dy_i \big|_E = 0 \right\}$$

Here $z_i = x_i + \sqrt{-1}y_i$ is the complex coordinate of $\mathbf{C}^n$. Then we have $\pi_1(Gr_{2n,n}) = \mathbf{Z}_2$, and its generator is the Stiefel-Whiteney class. On the other hand $\pi_1(Lag_n) = \mathbf{Z}$, and its generator is the universal Maslov class. The inclusion $Lag_n \to Gr_{2n,n}$ induces a surjection on their fundamental groups. This illustrates another way to explain the lifting of index from $\mathbf{Z}_2$ to $\mathbf{Z}$. (See Yoshida [Y].)

Now we show that our vector field $*F_a$ is a gradient vector field.

Let $a \in \mathcal{A}(N)$. We define its Chern-Simons functional by

$$cs(a) = \frac{1}{4\pi^2} \int_N \text{Tr}\left(\frac{1}{2} a \wedge da + \frac{1}{3} a \wedge a \wedge a \right).$$

Then, for $\mathcal{G}(N)$, we have

$$cs(g^*a) = cs(a) + \deg g.$$

Thus $cs: \mathcal{A}(N) \to \mathbf{R}$ induces a map : $\mathcal{B}(N) \to S^1 \cong \mathbf{R}/\mathbf{Z}$, which we denote also by cs.

We next fix a Riemannian metric on M, which induces metrics on $\mathcal{A}(N)$ and $\mathcal{B}(N)$. We then can define the gradient vector field of cs.

Lemma 4.1 (Taubes)   *The gradient vector field of* cs *is given by*

$$\frac{\partial a_t}{\partial t} = \frac{1}{4\pi^2} * F_{a_t}.$$

Proof.   We calculate

$$\frac{dcs(a_t)}{dt} = \frac{1}{4\pi^2}\frac{d}{dt}\int_N \text{Tr}\left(\frac{1}{2}a_t \wedge da_t + \frac{1}{3}a_t \wedge a_t \wedge a_t\right)$$

$$= \frac{1}{4\pi^2}\int_N \text{Tr}\left(\frac{1}{2}(\dot{a}_t \wedge da_t + a_t \wedge d\dot{a}_t) + \frac{1}{3}(\dot{a}_t \wedge a_t \wedge a_t + a_t \wedge \dot{a}_t \wedge a_t + a_t \wedge a_t \wedge \dot{a}_t)\right)$$

$$= \frac{1}{4\pi^2}\int_N \text{Tr}(\dot{a}_t \wedge da_t + \dot{a}_t \wedge a_t \wedge a_t)$$

$$= \frac{1}{4\pi^2}\int_N \langle \dot{a}_t, *(da_t + a_t \wedge a_t)\rangle \text{vol}N$$

The lemma follows.

Before going further, we recall Morse theory in the case of finite dimension and describe the procedure for finding homology group using Morse function.

Let $f: X \to \mathbf{R}$ be a function. We choose a metric on X. Put $Cr(f) = \{p \in X \mid df(p) = 0\}$. Let $\text{Hess}_p f: T_p X \to T_p X$ be the Hessian of f at p. The assumption that f is a Morse function means that $\text{Hess}_p f$ is invertible at each point $p \in Cr(f)$. We define the Morse index $\mu(p)$ by :

$$\mu(p) = \text{the number of negative eigenvalues of } \text{Hess}_p f.$$

We remark that $(-1)^{\mu(p)}$ is the index of the vector field $\text{grad } f$ at p.

Example 4.2   Put $X = S^2$ and let f be the function in Figure 4.3 :

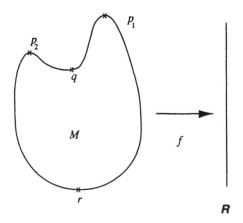

Figure 4.3

Then

$$\begin{cases} Cr(f) & = \{p_1, p_2, q, r\} \\ \mu(p_1) & = \mu(p_2) = 2, \\ \mu(q) & = 1 \\ \mu(r) & = 0 \end{cases}$$

For $p, q \in Cr(f)$, we put

$$\mathcal{M}(p,q) = \left\{ \ell : \mathbf{R} \to X \left| \begin{array}{l} \lim_{t \to \infty} \ell(t) = p, \\ \lim_{t \to -\infty} \ell(t) = q, \\ \dfrac{d\ell}{dt} = -\operatorname{grad} f \end{array} \right. \right\}.$$

We define an equivalence relation $\sim$ on it by $\ell \sim \ell_C$, $\ell_C(t) = \ell(t+C)$. Let $\overline{\mathcal{M}}(p,q) = \mathcal{M}(p,q) / \sim$. For the Morse function in Example 4.2, we have

$$\overline{\mathcal{M}}(p_1, q) = \overline{\mathcal{M}}(p_2, q) = \text{one point,}$$
$$\overline{\mathcal{M}}(q, r) = \text{two points,}$$

To find $\overline{\mathcal{M}}(p_i, r)$, we consider the figure 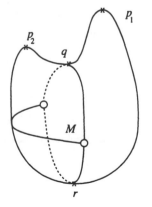 in the Figure 4.4 and remark that every element of $\mathcal{M}(p_1, r)$ intersects to it at exactly one point.

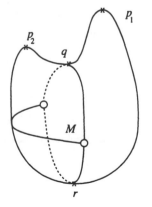

Figure 4.4

Hence

$$\overline{\mathcal{M}}(p_1, r) = (0,1) = \text{open interval}.$$

Similarly

$$\overline{\mathcal{M}}(p_2, r) = (0,1) = \text{open interval}.$$

We have :

Lemma 4.5 *If f is generic then*

$$\dim \mathcal{M}(p,q) = \mu(p) - \mu(q).$$

We put

$$
\begin{cases}
C_k(X,f) & = \displaystyle\bigoplus_{\substack{p \in Cr(f) \\ \mu(p)=k}} \mathbf{Z} \cdot [q] \\[1em]
\partial : & C_k(X,f) \to C_{k-1}(X,f) \\[1em]
\partial([p]) & = \displaystyle\sum_{\substack{q \in Cr(f) \\ \mu(q)=k-1}} \langle \partial p, q \rangle [q] \\[1em]
\langle \partial p, q \rangle & = \# \mathcal{M}(p,q).
\end{cases}
$$

Here # denotes the "order counted with sign", which we will explain below. There is a way of assigning orientations on each component of $\mathcal{M}(p,q)$. (In our case each of them is diffeomorphic to $\mathbf{R}$.) On the other hand, the group $\mathbf{R}$ acts on $\mathcal{M}(p,q)$ by $C \cdot \ell = \ell_C$. Hence (in the case $\dim \mathcal{M}(p,q) = 1$) there is another orientation on $\mathcal{M}(p,q)$ induced by the orientation of $\mathbf{R}$. Then $\# \mathcal{M}(p,q)$ is

The number of component of $\mathcal{M}(p,q)$ where the two orientations are coincide
    — the number of connected components where they are different.

A basic result is the following :

Theorem 4.6 (Morse-Thom-Smale-Witten [W1] - Floer etc.)

(4.6.1) $\partial \circ \partial = 0$

(4.6.2) $H_*(C_*(X, f)) = H_*(X, \mathbf{Z})$.

Let us first calculate the complex $C_*(X, f)$ for the Morse function in Example 4.2.

Since $\overline{\mathcal{M}}(p_1, q) = \overline{\mathcal{M}}(p_2, q) = $ one point, we have $\pm \partial[p_1] = \pm \partial[p_2] = [q]$. (Sign is not essential here.) On the other hand, $\overline{\mathcal{M}}(q, r)$ consistis of two points and if we examine their signs we can find that they cancel each other. Hence $\partial[q] = 0$. Thus we obtain

$$H_*(C_*(X, f)) = \begin{cases} [[p_1] + [p_2]] \cdot \mathbf{Z} & \text{if } k = 2, \\ 0 & \text{if } k \neq 0, 2, \\ [r] \cdot \mathbf{Z} & \text{if } k = 0, \end{cases}$$

for our Morse function f. This of course coincides with the homology of S^2.

Sketch of Proof of 4.6.1

We want to use the following lemma. Assume $\eta(p) - \eta(s) = 2$. (Then $\overline{\mathcal{M}}(p, s)$ is one dimensional. (Put $k = \eta(s) = \eta(p) - 2$.)

Lemma 4.7 $\overline{\mathcal{M}}(p, s)$ *is compactified such that*

$$\partial \overline{\mathcal{M}}(p, s) = \bigcup_{\eta(q) = k+1} \overline{\mathcal{M}}(p, q) \times \overline{\mathcal{M}}(q, s).$$

In place of proving lemma, let us looking at the case of Example 4.2. As we explained,

$\overline{\mathcal{M}}(p, s)$ =   = arc. The boundary of it consists of two points, which

corresponds to

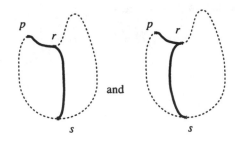

Thus $\partial\overline{\mathcal{M}}(p,q) = \overline{\mathcal{M}}(p,r) \times \overline{\mathcal{M}}(r,s)$. Here $\overline{\mathcal{M}}(p,r) =$ one point, $\overline{\mathcal{M}}(r,s) =$ two point. The lemma then follows.

By Lemma 4,7, we have

$$\sum_{\eta(q)=k-1} < \partial p, q >< \partial q, s >= 0$$

for each $p,s \in Cr(f)$, $\eta(p) = \eta(s) + 2$. It follows immediately that $\partial\partial = 0$.

Floer called $C_*(X,f)$ the *Witten complex.*

Now, we go back to the infinite dimensional situation we are studying. We put

$$X = \mathcal{B}(N,P),$$
$$f = cs.$$

We recall $\mathrm{grad}_a f = *F_a$, (up to constant). Then, for $[a],[b] \in Cr(cs) = R(N)$, we put

$$\mathcal{M}([a],[b]) = \left\{ [a_t] \left| \begin{array}{l} t \mapsto [a_t] \text{ is a path in } \mathcal{B}(N,P), \\ [a_{-\infty}] = [a], \quad [a_\infty] = [b], \\ \dfrac{da_t}{dt} = - * F_{a_t}. \end{array} \right. \right\}$$

In this infinite dimensional situation, Lemma 4.5 is modified as follows :

Theorem 4.8 (Floer [F]) *After a generic perturbation, we can find a map* $\mu: R(N) \to \mathbf{Z}_8$ *such that*

$$\dim \mathcal{M}([a],[b]) \equiv \mu(a) - \mu(b) \qquad \mod 8.$$

More precisely, each connected component of $\mathcal{M}([a],[b])$ is a smooth manifold whose dimension is congruent to $\mu(a) - \mu(b)$ modulo 8. (The dimension may depend on each component.)

Let us discuss a bit more about the ambiguity of the Morse index μ. First we recall

Fact 4.9 $\pi_1(\mathcal{B}(N,P)) \cong \mathbf{Z}$.

In fact, $\pi_1(\mathcal{B}(N,P)) \cong \pi_0(\mathcal{G}(N,P)) = \pi_0(\mathrm{Map}(N^3, SU(2)))$.
Therefore

$$\Omega(\mathcal{B}(M,P)) = \left\{ [a_t] \,\middle|\, \begin{array}{l} t \mapsto [a_t] \text{ is a path in } \mathcal{B}(N,P), \\ [a_{-\infty}] = [a], \quad [a_\infty] = [b]. \end{array} \right\}$$

has infinitely many components parametrized by $\pi_1(\mathcal{B}(N,P)) \cong \mathbf{Z}$. A better way to describe it is as follows. Consider the action of $\pi_1(\mathcal{B}(N,P))$ to $\pi_0(\Omega(\mathcal{B}(N,P)))$.

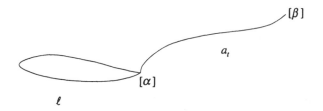

For $\alpha \in \pi_0(\Omega(\mathcal{B}(N,P)))$, let $\Omega_\alpha(\mathcal{B}(N,P))$ be the corresponding connected component of $\Omega(\mathcal{B}(M,P))$. Also, we put $\mathcal{M}_\alpha([a],[b]) = \mathcal{M}([a],[b]) \cap \Omega_\alpha(\mathcal{B}(M,P))$. Then the modulo 8 ambiguity of dimension of $\mathcal{M}([a],[b])$ can be described as follows.

Fact 4.10   *For* $k \in \mathbf{Z} = \pi_1(\mathcal{B}(N,P))$, $\alpha \in \pi_0(\Omega(\mathcal{B}(N,P)))$, *we have*

$$\dim \mathcal{M}_{k \cdot \alpha}([a],[b]) = 8k + \dim \mathcal{M}_\alpha([a],[b]).$$

(In fact, $k \in \mathbf{Z}$ is the difference of the relative Chern numbers of elements of $\mathcal{M}_{k \cdot \alpha}([a],[b])$ and $\mathcal{M}_\alpha([a],[b])$. (They are regarded as connections of $N \times \mathbf{R}$.) $8k$ is related to the bubbling phenomenon discussed in § 3.)

Let $\overline{\mathcal{M}}([a],[b])$ denote the quotient of $\mathcal{M}([a],[b])$ by the $\mathbf{R}$-action.

Now we define

Definition 4.11 (Floer [F1]) For $k \in \mathbf{Z}$, we put

$$CF_k(N) = \bigoplus_{\substack{[a] \in R(N) \\ \mu([a])=k}} \mathbf{Z} \cdot [a],$$

$$\begin{cases} \partial: \quad CF_k(N) \to CF_{k-1}(N) \\[2ex] \partial([a]) \qquad = \sum_{\substack{[b] \in R(N) \\ \mu([b])=k-1}} \langle \partial a, b \rangle [b] \\[2ex] \langle \partial a, b \rangle \qquad = \# \overline{\mathcal{M}}([a],[b]). \end{cases}$$

Theorem 4.12 (Floer) *Let N be a homology 3-sphere. Then*

(4.12.1) $\partial \circ \partial = 0$.

(4.12.2) $HF_*(N) \overset{def}{=} H_*(CF_*(N))$ *is independent of the choice of a metric and is an invariant of N.*

We call it the *Floer homology*. (Floer called $HF_*(N)$ the instanton homology.)

We finally explain a bit the assumption that N is a homology 3-sphere. In our situation, the space $\mathcal{B}(N,P)$ is in fact not a manifold but has a singularity. Hence, in general, there are several troubles for Morse theory. But, if N is a homology 3-sphere, one can prove that singularity does not cause serious trouble for the definition of $HF_*(N)$. This point will be explained a bit more in § 6.

§ 5   Donaldson invariant as topological field theory

In this section we employ the Floer homology to justify the discussion in § 1 about Donaldson invariant on 4-manifold with boundaries.

Before doing it, we discuss Morse theory in finite dimensions a bit more. We first give another example of Morse functions.

Example 5.1   Let $X = T^2 = \mathbf{R}^2 / \mathbf{Z}^2$ and $f : X \to \mathbf{R}$ be a function such that $f([x, y]) = \cos 2\pi x + \cos 2\pi y$. The vector field $-\text{grad } f$ is described as in Figure 5.2.

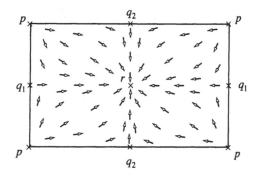

Figure 5.2

There are 4 critical points p, q_1, q_2, r which have the Morse index $2, 1, 1, 0$, respectively. It is easy to see that $\partial = 0$ in Witten's complex $C(X, f)$. Hence

$$H_k(C_*(X, f)) = \begin{cases} \mathbf{Z} & \text{if } k = 0, 2, \\ \mathbf{Z}^2 & \text{if } k = 1, \\ 0 & \text{otherwise.} \end{cases}$$

This (of course) coincides with the homology of the 2-torus, T^2.

Now we mention a few more properties of the Witten complex. First let us consider the Poincaré duality.

Let $f : X \to \mathbf{R}$ be a Morse function (on a finite dimensional manifold X) and $C(X, f)$ be its Witten complex. We put (as in the last section)

$$\mathcal{M}(p,q;f) = \left\{ \ell : \textbf{R} \to X \left| \begin{array}{l} \ell(-\infty) = p, \ell(+\infty) = q \\ \dfrac{d\ell}{dt} = -\text{grad } f \end{array} \right. \right\}.$$

Then it is easy to see that $Cr(f) = Cr(-f)$, $\mathcal{M}(p,q;f) = \mathcal{M}(q,p;-f)$. Moreover $\mu_f(p) = n - \mu_{-f}(p)$. Here $\mu_f(p)$, $\mu_{-f}(p)$ are the Morse indices with respect to the Morse functions $\pm f$ respectively, and n is the dimension of X. Hence the boundary operators of $C(X,f)$ are adjoint to those of $C(X,-f)$. Therefore

(5.3) $$H_k(C_*(X,f)) \cong H^{n-k}(C_*(X,-f)).$$

On the other hand, by (4.6.2) $H_k(C_*(X,f)) \cong H_k(C_*(X,-f)) \cong H_k(X;\textbf{Z})$. Hence we obtain the Poincaré duality.

Let us consider the case of Floer homology $HF(N^3)$.

We recall that we took $f = cs$. Since the Chern-Simons functional is defined by integration, if we change the orientation of the manifold N, the sign of $f = cs$ changes. We can prove that (5.3) holds as well in this infinite dimensional situation. Hence we have

$$"\,HF_k(N) = H_{\infty/2+k}(C_*(\mathcal{B}(N,P),cs)) \underset{(5.3)}{\cong} H_{\infty-(\infty/2+k)}(C_*(\mathcal{B}(N,P),-cs)) = HF^{-k}(-N).\,"$$

However, owing to a reason which we do not explain here, the degree shifts by 3. Hence we have

(5.4) $$HF_k(N) \cong HF^{3-k}(-N).$$

We remark that in our infinite dimensional situation there is no reason that $H_k(C_*(X,f)) \cong H_k(C_*(X,-f))$ holds. In fact, $HF_k(N) \neq HF_k(-N)$ in general.

For torsion free part, (5.4) can be rewritten as follows. We define

$$\bullet : CF_k(N) \otimes CF_{3-k}(-N) \to \textbf{Z},$$

by

$$[p] \bullet [q] = \begin{cases} 1 & \text{if } p = q, \\ 0 & \text{otherwise.} \end{cases}$$

Then

Fact 5.5   • *induces a unimodular perfect pairing between* $\dfrac{HF_k(N)}{Torsion}$ *and* $\dfrac{HF_{3-k}(-N)}{Torsion}$,

We return to the case of finite dimensional Morse functions $f: X \to \boldsymbol{R}$, and discuss the isomorphism (4.6.2), $H_*(C_*(X, f)) = H_*(X, \boldsymbol{Z})$. Let $i: H_*(C_*(X, f)) \to H_*(X, \boldsymbol{Z})$ be that isomorphism. To describe it, we define stable and unstable manifolds. Let $\varphi_t: X \to X$ be the one parameter group of transformations associated to the vector field grad f. Namely

$$\frac{d\varphi_t(x)}{dt} = \mathrm{grad}_{\varphi_t(x)} f,$$
$$\varphi_0(x) = x.$$

For $p \in Cr(f)$, we put

$$S(p) = \left\{ x \in X \;\middle|\; \lim_{t \to \infty} \varphi_t(x) = p \right\},$$
$$U(p) = \left\{ x \in X \;\middle|\; \lim_{t \to -\infty} \varphi_t(x) = p \right\}.$$

(We remark $\mathcal{M}(p, q) \underset{\text{diffeo}}{\cong} S(p) \cap U(q)$.)

Proposition 5.6   *If* $p \in Cr(f)$, $\partial[p] = 0$. *Then the closure* $\overline{U(p)}$ *is a cycle which represents* $i([p]) \in H_k(X; \boldsymbol{Z})$.

Let us now consider the case of the Morse function $f: X \to \boldsymbol{R}$ in Example 5.1. Consider $q_1 \in Cr(f)$ It satisfies $\partial[q_1] = 0$. Then $U(q_1)$ is as in Figure 5.7 below.

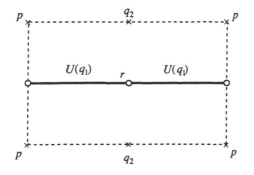

Figure 5.7

Hence $\overline{U(q_1)} \cong S^1$ is a cycle as Proposition 5.6 asserts.

On the other hand, let us consider the Morse function $f: X \to \textbf{R}$ in Example 4.2, and take $p_1 \in Cr(f)$. In this case, $\partial[p_1] = [q] \neq 0$. Hence the assumption of Proposition 5.6 is not satisfied. We find that $\overline{U(p_1)}$ is as in Figure 5.8 below.

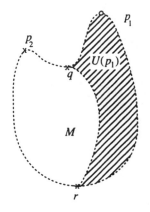

Figure 5.8

Then, in this case, $\overline{U(p_1)}$ is not a cycle. (Namely, its boundary is nonempty.)

In Proposition 5.6, we consider only the case when $\partial[p] = 0$. In general, if $\sum n_i \cdot \partial[p_i] = 0$, then $\sum n_i \cdot [\overline{U(p_i)}]$ is a cycle and represents $i\left(\left[\sum n_i \cdot [p_i]\right]\right)$.

Now we are in a position to describe the Donaldson invariant of 4-manifolds with boundary using the Floer homology.

We consider a 4-manifold M with boundary N and an $SU(2)$-bundle P on M. We put a metric on M so that M is a complete Riemannian manifold and that there exists a compact subset K for which the complement $M - K$ is isometric to the direct product $(0, \infty) \times N$. We are going to study the moduli space of ASD connections on M, that is,

$$\mathcal{M}(M, P) = \left\{ [A] \in \mathcal{B}(M, P) \mid *F_A = -F_A \right\}.$$

Let us consider the ASD equation

$$*F_A = -F_A,$$

on $(0,\infty) \times N \subset M$. Let t denote the coordinate in $(0,\infty)$ direction. We fix a trivialization of P on $(0,\infty) \times N$ and identify a connection on it with a section of $\Lambda^1((0,\infty) \times N) \otimes su(2)$. Hence a connection A is written as

$$A = a_t + b_t \wedge dt.$$

Here $t \mapsto a_t$ is a map $: (0,\infty) \to \Gamma(N, \Lambda^1(N))$, and $t \mapsto b_t$ is a map $(0,\infty) \to \Gamma(N, \Lambda^0(N))$.

Lemma 5.9   *There exists a gauge transformation $g \in (0,\infty) \times N \to SU(2)$ such that g^*A has no dt component.*

Proof.   We calculate

$$g^*(a_t + b_t \wedge dt) = g^{-1}a_t g + g^{-1}d_N g + g^{-1}\frac{dg}{dt} \wedge dt + g^{-1}b_t g \wedge dt.$$

Hence g^*A has no dt component if and only if

(5.10) $$\frac{dg}{dt} + b_t g = 0.$$

It is easy to see that (5.10) has always a solution.

By Lemma 5.9 we may assume $A = a_t$. We call this procedure "taking temporal gauge."

Now we calculate the ASD equation in temporal gauge. The following lemma plays an essential role in relating the ASD equation to the Floer homology.

Lemma 5.11 (Taubes)   *The connection $A = a_t$ is an ASD connection if and only if*

$$\frac{da_t}{dt} = *_N F_{a_t}.$$

Here $*_N$ is the Hodge star operator on the 3-manifold N, and F_{a_t} is the curvature of the connection a_t on N.

Proof.   The curvature of $A = a_t$ is calculated as

$$F_A \quad = d_M A + A \wedge A$$

$$= d_N a_t + a_t \wedge a_t - \frac{da_t}{dt} \wedge dt$$

Hence

$$*_M F_A \quad =*_M \left(d_N a_t + a_t \wedge a_t - \frac{da_t}{dt} \wedge dt \right)$$

$$= -*_N \left(d_N a_t + a_t \wedge a_t \right) \wedge dt +*_N \left(\frac{da_t}{dt} \right).$$

The lemma now follows.

Lemma 5.11 plays a key role in this section, since it relates the ASD equation to the gradient flow of the Chern-Simons functional. In fact, if we put $M = \mathbf{R} \times N$, then a connection $A = a_t$ is an ASD connection if and only if $t \mapsto [a_t]$, $\mathbf{R} \to \mathcal{B}(N,P)$ is a gradient line of the Chern-Simons functional.

Now, we recall the "definition" of the Donaldson invariant for 4-manifold with boundary, which we discussed in § 1. Here we take a metric on M (which has a boundary N) such that it is compact and that there is a neighborhood of N which is isometric to $(-T,0) \times N$. We write this manifold M_0 to distinguish it from the one with complete metric. (We may choose a metric so that $M = M_0 \cup_{N \times \{0\}} (0,\infty) \times N$.) We consider $\mathcal{M}(M_0,P) = \left\{ [A] \in \mathcal{B}(M_0,P) \mid *F_A = -F_A \right\}$, an $\infty/2$-dimensional submanifold of $\mathcal{B}(M_0,P)$. Also let $res: \mathcal{B}(M_0,P) \to \mathcal{B}(N,P)$ be the restriction map. Then, roughly, the Donaldson invariant is $res_*[\mathcal{M}(M_0,P)] \in H_{\infty/2+k}(\mathcal{B}(N,P))$. An equivalent way to describe it is the following. Let $[Z] \in H_{\infty/2+k}(\mathcal{B}(N,P))$. (Namely, Z is an $\infty/2$-dimensional cycle of $\mathcal{B}(N,P)$.) Then we consider the intersection $res^{-1}(Z) \cap \mathcal{M}(M_0,P)$. This is a finite (say k) dimensional cycle. Thus $[Z] \mapsto [res^{-1}(Z)] \bullet [\mathcal{M}(M_0,P)]$ is a Donaldson invariant.

To find such an $\infty/2$-dimensional cycle we recall Proposition 5.6. Suppose, for simplicity, $a \in R(N)$, $\partial[a] = 0$. Then we can take $Z = U(a)$, where $U(a)$ is the unstable manifold of the gradient flow of grad cs.

Warning : In fact, this unstable manifold is *not* well-defined.

Let us recall its definition. We "define" $\varphi_t: \mathcal{B}(N,P) \to \mathcal{B}(N,P)$ by

$$(5.12) \qquad \begin{cases} \dfrac{d\varphi_t(a)}{dt} & = -* F_{a_t}, \\ \varphi_0(a) & = a. \end{cases}$$

Warning : This map is *not* well-defined either, since (5.12) may not have a solution.

Then

$$(5.13) \qquad U([a]) \quad = \left\{ [b] \,\middle|\, \lim_{t \to \infty} [\varphi_t(b)] = [a] \right\}$$

$$= \left\{ [b] \,\middle|\, \begin{array}{l} \exists a_t : [0, \infty) \to \mathcal{A}(N, P), \\ a_0 = b, \quad [a_\infty] = [a], \\ \dfrac{da_t}{dt} = * F_{a_t}. \end{array} \right\}.$$

Then we "calculate"

$$(5.14) \qquad \mathcal{M}(M_0, P) \cap U([a]) \quad = \left\{ ([B],[b]) \,\middle|\, \begin{array}{l} [B] \in \mathcal{B}(M, P), \\ - F_B = * F_B, \\ [B|_N] = [b], \\ [b] \in U([a]). \end{array} \right\}$$

$$= \left\{ ([B],[A]) \,\middle|\, \begin{array}{l} [B] \in \mathcal{B}(M, P), \\ - F_B = * F_B, \\ A|_N = B|_N = b, \\ A = a_t, \text{ (no } dt \text{ component)}, \\ - F_A = * F_A, \\ A|_{\{\infty\} \times N} = a_\infty = a. \end{array} \right\}$$

By $A|_N = B|_N = b$, one can patch two connections A and B to obtain a connection C. Thus

$$(5.15) \qquad \mathcal{M}(M_0, P) \cap U([a]) \quad = \left\{ ([C]) \,\middle|\, \begin{array}{l} [C] \in \mathcal{B}(M, P), \\ - F_C = * F_C, \\ A|_{\{\infty\} \times N} = a. \end{array} \right\}.$$

The two formulas in (5.13), (5.14) do not make sense, since, for example, the unstable manifold is not well-defined. But we can prove that the right hand side of Formula (5.15) is well-defined and consists of finite dimensional manifold.   We put $\mathcal{M}(M,P;[a]) = $ The right hand side of (5.15).

Theorem 5.16 (Donaldson - Floer [DFK]) $\mathcal{M}(M,P;[a])$ *is a smooth manifold of dimension* $8k + 3(b_1 - 1 - b_2^+) + \eta(a)$.

"Theorem 5.16" bis   *If* $\partial^*([a]) = 0$, *then the closure* $\overline{\mathcal{M}(M,P;[a])}$ *is a cycle in* $\mathcal{B}(M,P)$. *If* $\sum n_i \cdot \partial^*[a_i] = 0$, *then* $\sum n_i \left[\overline{\mathcal{M}(M,P;[a_i])}\right]$ *is a cycle.*

We define $Q: HF^k(N;\mathbf{Q}) \to H_{8k+3(b_1-1-b_2^+)+k}(\mathcal{M}(M,P);\mathbf{Q})$ by $[a] \mapsto \left[\overline{\mathcal{M}(M,P;[a])}\right]$. To be slightly more precise, let us consider the map $\mu: H_k(M;\mathbf{Z}) \to H^{4-k}(\mathcal{B}(M,P);\mathbf{Z})$. Using this we define

Definition   5.17   We   define   relative Donaldson   invariant $Q: H_2(M;\mathbf{Z})^{\otimes \ell} \otimes HF^k(N;\mathbf{Q}) \to \mathbf{Q}$,   for $2\ell \equiv 3(b_1 - 1 - b_2^+)$,   by $Q(x_1, \cdots, x_\ell, [a]) = \overline{\mathcal{M}(M,P;[a])} \cap \prod \mu(x_i)$.

We may also regard $Q: H_2(M;\mathbf{Z})^{\otimes \ell} \to HF_k(N;\mathbf{Q})$.

Finally, we state the coupling formula in the case when N is a homology sphere. Suppose $\partial M_1 = N = -\partial M_2$. Put $M = M_1 \cup_N M_2$.
   Let
$$\begin{cases} \Sigma_i \in M_1, i = 1, \cdots, \ell \\ \Sigma_i \in M_2, i = \ell + 1, \cdots, \ell + \ell'. \end{cases}$$

Then $Q_{\ell+\ell'}(M; \Sigma_1, \cdots, \Sigma_{\ell+\ell'}) \in \mathbf{Q}$, $Q_\ell(M_1; \Sigma_1, \cdots, \Sigma_\ell) \in HF_k(N;\mathbf{Q})$, $Q_{\ell'}(M_2; \Sigma_{\ell+1}, \cdots, \Sigma_{\ell+\ell'}) \in HF^{3-k}(-N;\mathbf{Q}) \cong \left((HF_k(N;\mathbf{Q})\right)^*$.

Theorem 5.18 (Donaldson - Floer)
$$Q_{\ell+\ell'}(M; \Sigma_1, \cdots, \Sigma_{\ell+\ell'}) = \left\langle Q_{\ell'}(M_2; \Sigma_{\ell+1}, \cdots, \Sigma_{\ell+\ell'}), Q_\ell(M_1; \Sigma_1, \cdots, \Sigma_\ell) \right\rangle.$$

Let us mention a corollary of this theorem.

Theorem 5.19 (Donaldson [D4])   *Let* M *be a 4-manifold such that* $M = M_1 \# M_2$ *with* $b_2^+(M_1) > 0$, $b_2^+(M_2) > 0$. *Then the Donaldson invariant of* M *vanishes.*

§ 6   Gauge Theory on 4 manifolds with product end

In this and the next sections, we are going to make our argument a bit more precise to justify the heuristic argument in § 5. In § 2 we considered the ASD equation on closed 4-manifolds. But we need to study noncompact 4-manifolds here. As a consequence there will be various points at which we have to modify the argument in § 2. First, we need to find appropriate boundary condition to carry out functional analysis. Second, the noncompactness of 4-manifolds under consideration causes various troubles in the compactification of the moduli space. The difficulty of the problem depends on the 3-manifold which appears as the end of the moduli space or equivalently to what extent the ASD equation degenerates at infinity.

We begin with the simplest case and proceed to more and more general ones. In fact, the study of the most general case is still on progress.

It might be easier to explain these problems by using the (infinite dimensional) Morse theory picture which we introduced in the last section. So, let us consider a (closed) 3-manifold N and study the ASD equation on $N \times \textbf{R}$. As we explained in the previous section, this is equivalent to studying the gradient lines of the function cs on $\mathcal{B}(N,P)$. There are basically two troubles in analyzing the space of these gradient lines.

(A) cs **is in fact not a Morse function but is degenerate.**

(B) $\mathcal{B}(N,P)$ **is not smooth and has a singularity.**

These problems are related to each other. We first describe when they occur.

(B)   Singularity of $\mathcal{B}(N,P)$.

As in § 2^1, $[a] \in R(N)$ is a singular point if and only if the image of the holonomy $h_a : \pi_1(N) \to SU(2)$ is abelian. Moreover $I_a = \left\{ g \in \mathcal{G}(N,P) \mid g^*(a) = a \right\}$ is the centralizer of the image of h_a. We remark :

$$\frac{\text{Hom}(\pi_1(N), SU(2))}{\text{conjugate}} \neq \{1\}$$

if and only if $H^1(N; \textbf{Z}) \neq 0$.

Hence, in the case when N is a homology sphere, there is only one singularity on $R(N) = Cr(cs)$, that is, the trivial connections.

---

[1]There we considered 4-manifolds and here 3-manifolds. But the situation is the same for this problem.

The next simpler case is the case when N is a rational homology sphere[2]. In this case, there is only finitely many elements $[a] \in R(N)$ for which $I_a \neq \{\pm 1\}$. However, one remark that such elements may not be isolated in $R(N)$.

The other relatively simple case is the case when $H^1(N;\mathbf{Z})$ is torsion free (say isomorphic to $\mathbf{Z}^k$.) In this case, the moduli of elements $[a] \in R(N)$ for which $I_a \neq \{\pm 1\}$ is determined by a simple argument. Namely, the set of such an element is diffeomorphic to $T^k/\{\pm 1\}$. Where $T^k = \dfrac{\mathbf{R}^k}{\mathbf{Z}^k}$ and -1 acts on $T^k = \dfrac{\mathbf{R}^k}{\mathbf{Z}^k}$ by $[x_1,\cdots,x_k] \mapsto [-x_1,\cdots,-x_k]$. This space itself is singular and its singularity corresponds to the set where image of h_a is contained in the center of $SU(2)$. Namely, the set $H^1(N;\mathbf{Z}_2)$.

(A) When cs degenerates ?

As we·proved in the previous section, the Hessian of cs at $[a] \in R(N)$ is given by $*d_a : \operatorname{Ker} d_a^* \to \operatorname{Ker} d_a^*$. (Here $\operatorname{Ker} d_a^* = T_{[a]}\mathcal{B}(N,P) \subseteq \Gamma(N;\Lambda^1 \otimes AdP)$.) Hence $[a] \in R(N)$ is a nondegenerate critical point of cs if and only if

$$\operatorname{Ker} * d_a \cap \operatorname{Ker} d_a^* = \{0\}.$$

By harmonic theory, this condition is equivalent to

$$H^1(N;su(2)^a) = 0.$$

Here $su(2)^a$ denotes the local system corresponding to the composition of the holonomy representation h_a of $\pi_1(N)$ to $SU(2)$, and adjoint representation of $SU(2)$ to $su(2)$.

The simplest case.

Now we consider the moduli space :

$$\mathcal{M}(M,P;[a]) \quad = \quad \left\{ ([A]) \left| \begin{array}{l} [A] \in \mathcal{B}(M,P), \\ -F_A = *F_A, \\ A|_{\{\infty\}\times N} = a. \end{array} \right. \right\},$$

which we introduced at the end of last section. Here M is a 4-manifold having a compact

---

[2]Namely, $H^1(N;\mathbf{Q}) = 0$.

subset K such that $M - K$ is isometric to $N \times (0, \infty)$. We also consider

$$\mathcal{M}(M,P) \quad = \quad \left\{ ([A]) \left| \begin{array}{l} [A] \in \mathcal{B}(M,P), \\ -F_A = *F_A, \\ \int_M \|F_A\|^2 < \infty. \end{array} \right. \right\}.$$

The following two assumptions are related to the two troubles we mentioned above.

(1) $H^1(N; \mathbf{Z}) = 0$,

(2) $H^1(N; su(2)^a) = 0$ for arbitrary $[a] \in R(N)$.

Theorem 6.1 (Floer-Taubes) *Suppose N satisfies the above Assumptions (1) and (2). Then*

(1)
$$\mathcal{M}(M,P) \quad = \quad \bigcup_{[a] \in R(N)} \mathcal{M}(M,P;[a]).$$

Moreover, for each $[A] \in \mathcal{M}(M,P)$, there exists $B \in \mathcal{A}(M,P)$ which is gauge equivalent to A such that

(2)
$$\|B - a\| \leq Ce^{-ct}$$

on $M - K \cong (0, \infty) \times N$. Here t is the first coordinate and $[a]$ is an element of $R(N)$.

This theorem is an infinite dimensional analogue of the following result :

Fact 6.2   *Let X be a compact and finite dimensional manifold and $f : X \to \mathbf{R}$ be a Morse function. We put*

$$\mathcal{M}(X;f) = \left\{ \ell : \mathbf{R} \to X \left| \frac{d\ell}{dt} = -\mathrm{grad}\, f \right. \right\}.$$

Then, for each element $\ell \in \mathcal{M}(X;f)$, there exist $p, q \in Cr(f)$ such that $\ell \in \mathcal{M}(p,q)$. Moreover we have :

(6.3)
$$\left\{ \begin{array}{l} |\ell(t) - q| \leq Ce^{-ct} \\ |\ell(t) - p| \leq Ce^{ct} \end{array} \right. ,$$

for some positive constants C, c.

In place of proving Theorem 6.1, let us discuss a bit the proof of Fact 6.2. We assume that there exists a pair of points $p, q \in Cr(f)$ such that $\ell \in \mathcal{M}(p, q)$ and prove the Formula (6.3). We choose a coordinate of a neighborhood of q in such a way that $q = 0$ in this coordinate and that

$$f = \sum_{i=1}^{n} \lambda_i x_i^2.$$

Since f is a Morse function, it follows that $\lambda_i \neq 0$. We may assume that $\lambda_i > \lambda > 0$ for $i < m$ and $\lambda_i < -\lambda < 0$ for $i \geq m$. We may also assume that the Riemannian metric satisfies

$$g_{ij} = \delta_{ij} + o(|x|).$$

Hence the equation $\dfrac{d\ell}{dt} = -\text{grad } f$ is in our coordinate

$$(6.4) \qquad \frac{d\ell_i}{dt} = -\lambda_i \ell_i + \tau_i(\ell).$$

Here $\tau_i(\ell)$ satisfies the inequality

$$(6.5) \qquad |\tau_i(\ell)| < |\ell|^2.$$

Now, using the assumption $\lambda_i \neq 0$, it is rather elementary to prove that the nonlinear term does not change the asymptotic behavior of the solution. Hence we can forget it and consider the linearized equation

$$\frac{d\ell}{dt} = -\lambda_i \ell_i.$$

The solution of it is $\ell_i = a_i e^{-\lambda_i t}$. Then using the assumption $\lim_{t \to \infty} \ell_i(t) = 0$ we find that $a_i = 0$ for $i \geq m$. Thus we have

$$|\ell_i| \leq C e^{-ct},$$

as required.

Now we remark that Theorem 6.1 is a generalization of Theorem 2.5 in § 2. In fact, let A be an ASD $SU(2)$-connection on $D^4 - \{0\}$ such that $\int \|F_A\|^2 < \infty$. We recall that $D^4 - \{0\}$ is conformally equivalent to $S^3 \times (-\infty, 0]$. We also remark the important fact that the ASD equation and the L^2-norm of 2-forms are conformally invariant. Therefore the connection A will be transformed to an ASD connection $\hat{A}$ on $S^3 \times (-\infty, 0]$ such that $\int \|F_{\hat{A}}\|^2 < \infty$. Hence one can apply Theorem 6.1 to $\hat{A}$. We remark that there is only one flat connection on S^3, the trivial connection. Therefore, Theorem 6.2 implies that $\hat{A}$ is gauge equivalent to the connection $\hat{A}'$ such that $|\hat{A}'(t)| \leq Ce^{ct}$. One can use this inequality to show that $\hat{A}'$ is transformed to a connection on $D^4 - \{0\}$ which can be extended to a smooth connection on D^4. This is the conclusion of Theorem 2.5.

We next discuss the compactification of the moduli space $\mathcal{M}(M,P;[a])$. Roughly speaking, there are·two kinds of ends in this moduli space. One is related to the bubbling phenomenon. Namely, a sequence $A_i \in \mathcal{M}(M,P;[a])$ for which the norm $|A_i|$ diverges at some point in M. This phenomenon appeared in the case when M is compact (see Theorem 2.5), and basically can be handled in the same way as we explained in § 2.

There is another tye of ends in the case when M is noncompact. This also appeared in finite dimensional Morse theory, namely, in that case a gradient line splits into two gradient lines. Roughly speaking, this end corresponds to $\mathcal{M}(M,P;[b]) \times \overline{\mathcal{M}}([b],[a])$. (See Lemma 4.7.)

In this situation, however, there is one more factor to make our situation more complicated. That is the presence of trivial connection $[0]$. (We remark that the space $\mathcal{B}(N,P)$ is singular at $[0]$.) Here we describe how this singularity is related to the compactification of the moduli spaces.

To state the result, we first discuss the boundary condition of the gauge transformation. We put

$$\mathcal{A}(M,P;a) \qquad = \qquad \left\{ A \left| \begin{array}{l} A \in \mathcal{A}(M,P), \\ A|_{\{\infty\} \times N} = a. \end{array} \right. \right\},$$

Take an element g of $\mathcal{G}(M,P)$, the set of bundle automorphism of (M,P). (We did not assume boundary any condition in the definition of $\mathcal{G}(M,P)$.) Then g preserves the set $\mathcal{A}(M,P;a)$ if and only if g converges to an element of I_a as t goes to infinity. (Here we recall that I_a is the set of all gauge transformations on N which fixes a.) So we put

$$G(M,P;a) = \left\{ A \left| \begin{array}{l} g \in G(M,P), \\ g|_{\{\infty\} \times N} \in I_a. \end{array} \right. \right\}$$

$$G_0(M,P) = \left\{ A \left| \begin{array}{l} g \in G(M,P), \\ g|_{\{\infty\} \times N} \in \{\pm 1\}. \end{array} \right. \right\}.$$

We remark that if $[a]$ is nonsingular in $B(N,P)$, then these two groups coincide. Hence, in our situation, there is only one $[a]$ for which $G(M,P;a) \neq G_0(M,P)$, namely $[a] = [0]$. We put

$$\tilde{M}(M,P;a) = \left\{ [A] \left| \begin{array}{l} A \in \mathcal{A}(M,P;a), \\ -F_A = *F_A, \end{array} \right. \right\},$$

$$\hat{M}(M,P;a) = \tilde{M}(M,P;a)/G_0(M,P)$$

$$M(M,P;a) = \tilde{M}(M,P;a)/G(M,P).$$

Then $\hat{M}(M,P;a)/I_a = M(M,P;a)$ in general. (We remark that $I_a \subseteq SU(2)$.)

We next explain the relation between relative Chern numbers and the mod 8 ambiguity of the indices in Floer theory.

We first fix a trivialization of P at N, and hence the relative Chern number of P. If we change the trivialization by a gauge transform contained in the connected component $G_{conn}(N;P)$ of the gauge group, the relative Chern number do not change. So, if we take the equivalence class $[a]$ of a flat connection a in $\mathcal{A}(N,P)/G_{conn}(N;P)$, then the relative Chern number of the elements of $M(N,P;[a])$ is well-defined. Also if we remark the isomorphism $\pi_0(G(N,P)) \cong \pi_1(M(N,P))$ one finds that the Morse index of the equivalence class $[a]$ of a flat connection in $\mathcal{A}(N,P)/G_{conn}(N;P)$ is well-defined as an integer. (In other words the Chern-Simons functional is globally well-defined as a function from $\mathcal{A}(N,P)/G_{conn}(N;P)$ to $\boldsymbol{R}$.) Hereafter in this section, we regard $[a]$ as an element in $\mathcal{A}(N,P)/G_{conn}(N;P)$. Hence $\eta([a]) \in \boldsymbol{Z}$ and the relative Chern number of P is a well-defined integer.

If we fix the relative Chern number of P, then for each $g \in G_{conn}(M;P)$, we have $M(M,P;a) \cong M(M,P;g^*a)$. Thus we write $M(M,P;[a])$ in place of $M(M,P;a)$.

Now we describe the compactification as follows.

Theorem 6.5 (Probably in [DFK]) *There exists a space* $CM(M,P;[a])$ *such that*

(1)   *the complement of* $M(M,P;[a])$ *in* $CM(M,P;[a])$ *is a codimension one subset.*

(2)   *the complement of the union*

$$\bigcup_{[b]\in R(N)} \hat{\mathcal{M}}(M,P;[b])\times_{I_b} (\hat{\mathcal{M}}([b],[a])/\mathbf{R})$$

in $\mathcal{CM}(M,P;a)-\mathcal{M}(M,P;a)$ *is a set of* codim ≥ 2 *in* $\mathcal{CM}(M,P;[a])$.

(3)   *If* $[A_i]\in \mathcal{M}(M,P;[a])$ *is a sequence such that* $\sup|F_{A_i}|$ *is bounded uniformly for* i, *then* $[A_i]$ *has a subsequence converging on* $\mathcal{CM}(M,P;[a])$.

Roughly speaking Theorem 6.5 says that $\mathcal{CM}(M,P;[a])$ is a compactification of $\mathcal{M}(M,P;[a])$ modulo bubbling phenomenon. Moreover, its boundary is equal to the union of the spaces $\hat{\mathcal{M}}(M,P;[b])\times_{I_b} (\hat{\mathcal{M}}([b],[a])/\mathbf{R})$.

The space $\hat{\mathcal{M}}(M,P;[b])\times_{I_b} (\hat{\mathcal{M}}([b],[a])/\mathbf{R})$ coincides with $\mathcal{M}(M,P;[b])\times \overline{\mathcal{M}}([b],[a])$ if $I_b=\{\pm 1\}$, namely if $b\neq 0$. In case $b=0$, $\hat{\mathcal{M}}(M,P;[b])\times_{I_b} (\hat{\mathcal{M}}([b],[a])/\mathbf{R})$ coincides with an $SO(3)$-bundle over $\mathcal{M}(M,P;[0])\times \overline{\mathcal{M}}([0],[a])$.

Let us discuss a bit how this extra boundary $\bigcup_{[b]\in R(N)} \hat{\mathcal{M}}(M,P;[b])\times_{I_b} (\hat{\mathcal{M}}([b],[a])/\mathbf{R})$ arises. Let $[A_i]\in \mathcal{M}(M,P;a)$ be as in Theorem 6.5 (1). By using the argument in § 2, we can prove that, by taking a subsequence if necessary, there is a sequence of gauge transformations $g_i\in \mathcal{G}(M,P)$ such that $g_i^*(A_i)$ converges to A_∞ in the compact uniform topology. (We remark that no bubbling can occur, since we assume that $\sup|F_{A_i}|$ is bounded uniformly for i.) In case M is compact, it will imply that $g_i^*(A_i)$ converges uniformly and hence we are done. However, we are interested in the case when M is noncompact. We consider

$$C_i = \sup|F_{A_\infty}-F_{A_i}|.$$

If C_i converges to zero, then $g_i^*(A_i)$ converges uniformly and hence we are done.

Otherwise, we can find $p_i\in M$ such that $C_i=\left|F_{A_\infty}-F_{A_i}\right|(p_i)$ is bounded away from 0. Since $g_i^*(A_i)$ converges to A_∞ in the compact uniform topology, it follows that $p_i\in M$ diverges. So we may put $p_i=(t_i,x_i)\in [0,\infty)\times N$ with $\lim_{i\to\infty} t_i=\infty$. We take $s_i \ll t_i$ with $\lim_{i\to\infty} s_i=\infty$ and take an embedding $I_i:(-s_i,s_i)\times N\to M$, $(t,x)\mapsto (t+t_i,x)$. Then, after taking a subsequence, the sequence of ASD connections $g_i'^* I_i^* A_i$ converges to an ASD connection $A_{\infty,1}$, for some sequence of gauge transformations g_i'. Here, $A_{\infty,1}$ is a connection on $\mathbf{R}\times N$ and the convergence is taken by the compact uniform topology.

Since $\left\|F_{A_{\infty,1}}\right\|_{L^2}$ is smaller than $\limsup\left\|F_{A_i}\right\|_{L^2}$, it follows that $\left\|F_{A_{\infty,1}}\right\|_{L^2}$ is finite. Hence by Theorem 6.1, $[A_{\infty,1}] \in \mathcal{M}([b],[c])$ for some $[b],[c] \in R(M)$.

If $\left| F_{A_i}(t,x) - (F_{A_\infty}(x) + F_{A_{\infty,1}}(t+t_i,x)) \right|$ converges to 0 uniformly for i, then we can prove $[A_{\infty,1}] \in \mathcal{M}([b],[a])$ and $[A_\infty] \in \overline{\mathcal{M}}(M,P;[b])$, and define $\lim_{i \to \infty}[A_i] = ([A_\infty],[A_{\infty,1}])$. (See Figure 6.6.)

In this way one finds a boundary parametrized by $\mathcal{M}(M,P;[b]) \times \overline{\mathcal{M}}([b],[a])$. In the case when $b = 0$, extra parameter $SO(3)$ arises. We omit the discussion on it.

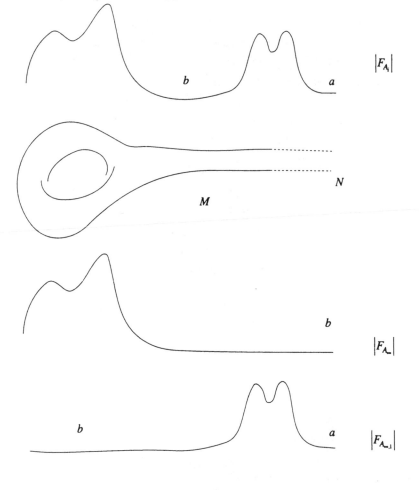

Figure 6.6

If $\left| F_{A_i}(t,x) - (F_{A_\infty}(x) + F_{A_{\infty,1}}(t+t_i,x)) \right|$ does not converge to 0, we can repeat the argument and find $[A_{\infty,2}]$.   In this way one finds $[A_{\infty,3}],\cdots,[A_{\infty,k}]$ such that $\left| F_{A_i}(t,x) - (F_{A_\infty}(x) + F_{A_{\infty,1}}(t+t_{i,1},x) + F_{A_{\infty,2}}(t+t_{i,2},x) + \cdots + F_{A_{\infty,k}}(t+t_{i,k},x)) \right|$ converges to 0. So this sequence corresponds to an element of

$$\bigcup_{[b_i] \in R(N)} \hat{\mathcal{M}}(M,P;[b_1]) \times_{l_{b_1}} (\hat{\mathcal{M}}([b_1],[b_2])/\boldsymbol{R}) \times_{l_{b_2}} \cdots \times_{l_{b_{k-1}}} (\hat{\mathcal{M}}([b_{k-1}],[a])/\boldsymbol{R}).$$

By calculating the dimension of it one finds that it is smaller than the dimension of $\bigcup_{[b] \in R(N)} \hat{\mathcal{M}}(M,P;[b]) \times_{l_b} (\hat{\mathcal{M}}([b],[a])/\boldsymbol{R})$. The assertion (2) holds.

The ends described in Theorem 6.5 is called the sliding ends.

In the case when $M = \boldsymbol{R} \times N$, Theorem 6.5 implies that

Corollary 6.7 (Floer [F])   *There exists a space* $C\overline{\mathcal{M}}([a],[b])$ *such that*

(1)   *the complement of* $\overline{\mathcal{M}}([a],[b])$ *in* $C\overline{\mathcal{M}}([a],[b])$ *is a codimension one subset.*

(2)   *the complement of the union of*

$$\bigcup_{\substack{[c] \in R(N) \\ c \neq 0}} \overline{\mathcal{M}}(M,[a];[b]) \times \overline{\mathcal{M}}([c],[a]),$$

and an $SO(3)$-*bundle over*

$$\overline{\mathcal{M}}(M,[a];[0]) \times \overline{\mathcal{M}}([0],[a])$$

in $C\overline{\mathcal{M}}([a],[b]) - \mathcal{M}(M,P;a)$ *is a set of* codim ≥ 2 *in* $C\overline{\mathcal{M}}([a],[b])$.

(3)   *If* $\eta(a) - \eta(b) < 8$, *then* $C\overline{\mathcal{M}}([a],[b])$ *is compact.*

We remark that Corollary 6.7 is almost the same as Lemma 4.7. The difference appears in the two points : the alternative end, which is an $SO(3)$-bundle over $\overline{\mathcal{M}}(M,[a];[0]) \times \overline{\mathcal{M}}([0],[a])$: the bubbling end which arises in case when $\eta(a) - \eta(b) \geq 8$. The first point is the influence of the singularity [0]. On the other hand, the second point is an infinite dimensional phenomenon which does not have an analogy in finite dimensional Morse theory. But when we examine the situation carefully we find that these two differences do not cause serious troubles to imitate the proof of Theorem 4.6 and to show Theorem 4.12.

In order to justify the coupling formula Theorem 5.18, one needs to study the moduli of ASD connections for the manifold $M = M_1 \cup_N M_2$.

We suppose that $M_1 - K_1 \cong (0, \infty) \times N$ and $M_2 - K_2 \cong (-\infty, 0) \times N$, for some compact sets K_i. We put $M_T = (K_1 \cup (0, T] \times N) \cup (K_2 \cup (-T, 0] \times N)$. ($M_T$ is $M = M_1 \cup_N M_2$ equipped with a Riemannian metric which depends on T.)

Now, roughly speaking, we can prove that

$$(6.8) \qquad \lim_{T \to \infty} \mathcal{M}(M_T, P) = \bigcup_{[a] \in R(N)} \mathcal{M}(M_1, P_1; [a]) \times \mathcal{M}(M_2, P_2; [a]).$$

(See Figure 6.10.) In this way one can find a relation between the moduli space of ASD connections of $M = M_1 \cup_N M_2$ and that on M_1, M_2. But (6.8) is not rigorous, since we need to handle trivial connection and also we have to make it clear what is meant by the limit of the moduli spaces. So we state the results as follows :

Theorem 6.9   *There exists a space $CM_{para}(M, P)$ such that*

(1)   *The complement of $\bigcup_{T \geq T_0} \mathcal{M}(M_T, P)$ in $CM_{para}(M, P)$ is a codimension one subset.*

(2)   *The complement of the union of $\mathcal{M}(M_{T_0}, P)$,*

$$\bigcup_{\substack{[a] \in R(N) \\ a \neq 0}} \mathcal{M}(M_1, P_1; [a]) \times \mathcal{M}(M_2, P_2; [a])$$

and

$$\bigcup \hat{\mathcal{M}}(M_1, P_1; [0]) \times_{SO(3)} \hat{\mathcal{M}}(M_2, P_2; [0])$$

in $CM_{para}(M, P) - \bigcup_{T \geq T_0} \mathcal{M}(M_T, P)$ is codim ≥ 2 in $CM_{para}(M, P)$.

(3)   *If $[A_i] \in \mathcal{M}(M_{T_i}, P)$ is a sequence such that $\sup |F_{A_i}|$ is bounded uniformly for i, then $[A_i]$ has a subsequence converging in $CM_{para}(M, P)$.*

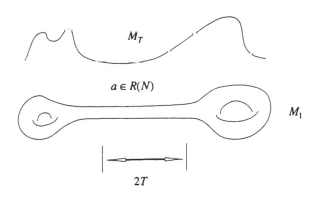

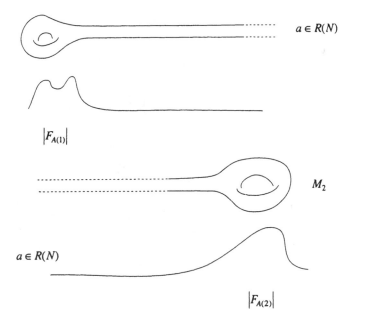

Figure 6.10

Roughly speaking, Theorem 6.9 says that $[\mathcal{M}(M_{T_0}, P)]$ is homologous to

$$\left[\bigcup_{\substack{[a] \in R(N) \\ a \neq 0}} \mathcal{M}(M_1, P_1; [a]) \times \mathcal{M}(M_2, P_2; [a])\right] + \left[\bigcup \hat{\mathcal{M}}(M_1, P_1; [0]) \times_{SO(3)} \hat{\mathcal{M}}(M_2, P_2; [0])\right].$$

Thus we described some of the results one needs to justify the construction in § 5, in the case when N is a homology sphere.

Now we discuss how to use Theorem 6.5 to justify the definition of the relative Donaldson invariant. We have a dimension formula

$$\dim \mathcal{M}(M,P;[a]) = (8c^2(P) \cap [M]) + \eta(a) - 3b_2^+.$$

(We assume that $H_1(M;\mathbf{Q}) = H_1(N;\mathbf{Q}) = 0$.)

Then we consider the case when the right hand side of the formula is 2ℓ. Namely we put $2\ell - (8c^2(P) \cap [M]) + 3b_2^+ = k$ and assume $\eta(a) = k$. Choose ℓ surfaces $\Sigma_1, \cdots, \Sigma_\ell$ in M. Then we defined in § 2 the complex line bundle $\mathcal{L}(\Sigma_i)$ on $\mathcal{B}(\Sigma_i, P)$. We choose a section s_Σ to $\mathcal{L}(\Sigma_i)$. Let $res_\Sigma : \mathcal{B}(M,P) \to \mathcal{B}(\Sigma_i, P)$ be the restriction map. Then, as in (2.8), we define

$$Q_\ell(\Sigma_1, \cdots, \Sigma_\ell; [a]) = [\mathcal{M}(M,P;[a])] \cap \bigcap_{i=1}^\ell [res_{\Sigma_i}^{-1}(s_{\Sigma_i}^{-1}(0))] \in \mathbf{Z}.$$

Now we have :

Theorem 6.11 $\displaystyle\sum_{\eta(a)=k} Q_\ell(\Sigma_1, \cdots, \Sigma_\ell; [a]) \cdot [a]$ *is a cycle.*

Sketch of the Proof

We have to prove

(6.12) $$\sum_{\eta(a)=k} Q_\ell(\Sigma_1, \cdots, \Sigma_\ell; [a]) < \partial a, b >= 0$$

for each b with $\mathcal{B}(\Sigma_i, P)$.

We remark that our surfaces Σ_i are disjoint with the boundary N. Hence one can prove that the map on $res_\Sigma : \mathcal{M}(M,P;[a]) \to \mathcal{B}(\Sigma_i, P)$ and $res_\Sigma : \mathcal{M}(M,P;[b]) \times_{l_b} (\hat{\mathcal{M}}([b],[a])/\mathbf{R}) \to \mathcal{B}(\Sigma_i, P)$ can be patched together to define a map on the compactification $res_\Sigma : C\mathcal{M}(M,P;[a]) \to \mathcal{B}(\Sigma_i, P)$. (Here the map $res_\Sigma : \mathcal{M}(M,P;[b]) \times_{l_b} (\hat{\mathcal{M}}([b],[a])/\mathbf{R}) \to \mathcal{B}(\Sigma_i, P)$ is constant in the second factor.)

Therefore, using Theorem 6.5, we can prove that $C\mathcal{M}(M,P;[b]) \cap \bigcap_{i=1}^\ell res_{\Sigma_i}^{-1}(s_{\Sigma_i}^{-1}(0))$ is a (one dimensional) manifold whose boundary is

$$(6.13) \qquad \bigcup_{\eta(u)=k} \left(\mathcal{M}(M,P;[a]) \cap \bigcap_{i=1}^{\ell} res_{\Sigma_i}^{-1}(s_{\Sigma_i}^{-1}(0)) \right) \times \overline{\mathcal{M}}([a],[h])$$

Since the left hand side of (6.12) is the order counted with sign of (6.13), it follows that it vanishes.

In fact, we need to see carefully the contribution of the trivial connection in order to make the above argument rigorous. The argument one need for it is described in detail in [D4]. (There the case $N = S^3$ is discussed. But the same argument can apply to the case when N is a homology sphere.)

We can use Theorem 6.10 to prove Theorem 5.18 in a similar way.

§ 7   Equivariant Floer theory, Higher boundary and degeneration at infinity

To study the Floer homology and the relative Donaldson invariant in general case, there are various factors we have to consider and the final result is not yet obtained. So we mention here some arguments and ideas which are to be building blocks of general results.

Equivariant (co)-homology.

As we mentioned before, the singularity of $\mathcal{B}(N,P)$ is one of the major sources of trouble. One can study this singularity using the technique of equivariant homology and cohomology. This idea was first proposed by D. Austin and P.Braam [AB 1,2]. C. Taubes [T6] is using the same kind of ideas. According to P. Braam, this ideas was implicit in Witten's paper [W2].

We fix $p_0 \in N$ and put $\mathcal{G}_0(N,P) = \left\{ g \in \mathcal{G}(N,P) \mid g(p_0) = id \right\}$. The action of $\mathcal{G}_0(N,P)$ is free on $\mathcal{A}(N,P)$. Hence the quotient space $\tilde{\mathcal{B}}(N,P)$ is smooth. On $\tilde{\mathcal{B}}(N,P)$, the group $SU(2) = \mathcal{G}(N,P) / \mathcal{G}_0(N,P)$ acts in such a way that the quotient space is $\mathcal{B}(N,P)$

The idea is to "define" $H_{SU(2)}^{\infty/2+*}(\tilde{\mathcal{B}}(N,P);\mathbf{Z})$ and use it. (It might be better to use $H_{SO(3)}^{\infty/2+*}(\tilde{\mathcal{B}}(N,P);\mathbf{Z})$.)

Let us recall the definition of equivariant cohomology in usual situation. Let M be a manifold on which a compact Lie group G acts from right. Take a contractible space EG on which G acts freely from left. Then we define

$$H_G^k(X;\mathbf{Z}) = H^k(X \times_G EG;\mathbf{Z}).$$

This group is a module over $H^*(BG;\mathbf{Z}) = H^*(EG/G;\mathbf{Z})$. Now we describe a way to find this group by using a Morse theory. Austin-Braam did it in [AB1], where they used a kind of mixture of the De-Rham theory and Morse theory. Taubes' and Witten's approach also uses De-Rham theory. Hence they works over $\mathbf{R}$. Here we discuss it in a bit different way based on "singular" homology. (In the situation of symplectic geometry Floer homology, where the group G is S^1, the same kind of idea is described by Floer [F2].

First of all, we need to find a "Morse function" on M which is G invariant. But such a function may not exists in general, because the critical point set $Cr(f)$ of a G invariant function is necessary G invariant and hence if, for example, G has no fixed points, it is impossible that $Cr(f)$ is discrete.   Therefore we need to generalize the situation to Bott-Morse functions.

Definition 7.1   A function $f:M \to \textbf{R}$ is called a *Bott-Morse function* if $Cr(f)$ is a smooth submanifold of M and if for each $p \in Cr(f)$ the restriction of $Hess_p f$ to the normal bundle of $Cr(f)$ is nondegenerate.

Example 7.2 :   Let us take $M = T^2$ and choose $f:M \to \textbf{R}$ as in Figure 6.13, where the height function is $f:M \to \textbf{R}$. Then $Cr(f)$ is $\{p_1, p_2, q_1, q_2\} \cup S^1$, and f is a Bott-Morse function.

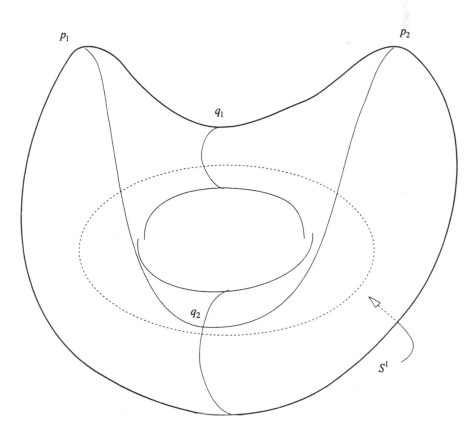

Figure 7.3

Let $Cr(f) = \bigcup_{a \in A} R_a$ be its decomposition to the connected components. By assumption, the number of negative eigenvalues of $Hess_p f$ is independent of the choice of $p \in R_a$ and depends only on a. We call it the Morse index of R_a and write $\eta(a)$. We imitate the construction in § 4 and put

$$\mathcal{M}(a,b) = \left\{ \ell : \boldsymbol{R} \to M \left| \begin{array}{l} \ell(-\infty) \in R_a, \\ \ell(\infty) \in R_b, \\ \dfrac{d\ell}{dt} = -\text{grad } f \end{array} \right. \right\}.$$

Lemma 7.4   *We can perturb the function f outside the critical point set in such a way that $\mathcal{M}(a,b)$ is a smooth manifold of dimension*

$$\eta(a) - \eta(b) + \dim R_a.$$

We divide $\mathcal{M}(a,b)$ by $\boldsymbol{R}$ action and let $\overline{\mathcal{M}}(a,b)$ be the quotient space.

For the Morse function in Example 7.2, we have

$$\begin{array}{lcl} \overline{\mathcal{M}}(p_i,q_j) & = & \text{one point,} \\ \overline{\mathcal{M}}(q_i,S^1) & = & \text{two points,} \\ \overline{\mathcal{M}}(p_i,S^1) & = & \text{two arcs.} \end{array}$$

Here we remark that in the case of Morse functions the space $\overline{\mathcal{M}}(p,q)$ is empty if $\eta(p) - \eta(q)$ is negative or if $\eta(p) - \eta(q) = 0, p \neq q$. In this case the result follows from Lemma 7.4 or Lemma 4.5. But, in the Bott-Morse situation the dimension formula has an additional term, $\dim R_a$. As a consequence there is no reason that $\mathcal{M}(a,b)$ is empty for each a,b such that $\eta(a) - \eta(b)$ is negative or $\eta(a) - \eta(b) = 0, a \neq b$. According to Austin-Braam [AB 1], we call the Bott-Morse function is *weakly self-indexing* if $\mathcal{M}(a,b)$ is empty for each a,b such that $\eta(a) - \eta(b)$ is negative or $\eta(a) - \eta(b) = 0, a \neq b$. Now we have

Theorem 7.5 ([AB 1],[Fk2])   *Let M be a closed manifold and $f:M \to \boldsymbol{R}$ be a weakly self-indexing Bott-Morse function. We put $R_i = \bigcup_{\eta(a)=i} R_a$. Then there exists a spectral sequence $E_{i,j}^k$ such that*

(1) $E_{i,j}^k$ converges to $H_{i+j}(M,\mathbf{Z})$,

(2) $E_{i,j}^1 \cong H_j(R_i;\mathbf{Z})$.

The differential of this spectral sequence is given as follows. We remark that there are maps $\pi_l:\overline{\mathcal{M}}(a,b) \to R_a$, $[\ell] \mapsto \ell(-\infty)$, and $\pi_r:\overline{\mathcal{M}}(a,b) \to R_b$, $[\ell] \mapsto \ell(\infty)$. We consider the diagram :

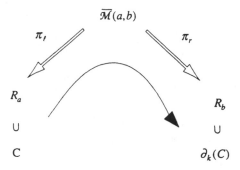

Then for an *n*- cycle $C \subset R_a \subset R_i$, the component in R_b of $\partial_k([C])$ (which is an $n+k-1$ - cycle in R_{i-k}), is represented by $\pi_r\left(\pi_l^{-1}(C)\right)$. The reader may try to calculate this spectral sequence in the case of Example 7.2. In fact, we need to modify this argument a bit, since the space $\overline{\mathcal{M}}(p,q)$ has a boundary in general.

Now, let us go back to the study of equivariant Morse theory. Let *M* be a closed manifold on which a compact group *G* acts smoothly. An important remark (see Austin-Braam [AB1]) is that we can *not* choose, in general, a *G*-invariant Bott-Morse function satisfying the conclusion of Lemma 7.4 by perturbation.

Example 7.6   We consider the function

$$f(x,y,z) = x^3 - 3x - \varepsilon x(y^2 + z^2).$$

This function is a Morse function and is invariant by an S^1-action, that is the rotation around *x*-axis. There is two critical points $(\pm 1,0,0)$ of *f*. The Morse index at $(1,0,0)$ is 2 and that at $(-1,0,0)$ is 1. On the other hand, there is a gradient line from $(-1,0,0)$ to $(1,0,0)$ which is contained in the *x*-axis. Hence $\mathcal{M}((-1,0,0),(1,0,0))$ is nonempty. However

$\eta(-1,0,0) - \eta(1,0,0) = -1$.

Suppose that f' is an S^1-invariant perturbation of f. Then x-axis is invariant by the gradient flow of f'. Hence one finds that there is again a gradient line of f' from $(-1,0,0)$ to $(1,0,0)$. Thus the moduli space $\mathcal{M}((-1,0,0),(1,0,0))$, whose virtual dimension is -1, can not be perturbed to be empty.

One way to explain this phenomenon is the following. Let $\ell \in \mathcal{M}((-1,0,0),(1,0,0))$. We consider the linearization of the equation $\dfrac{d\ell}{dt} = -\text{grad } f$. It defines an elliptic complex

$$(7.7) \qquad\qquad \frac{\partial}{\partial t} + \nabla\text{grad } f : \Gamma(\ell^* TM) \to \Gamma(\ell^* TM),$$

The index of this operator on $L^2(\ell^* TM)$ is $\eta(-1,0,0) - \eta(1,0,0)$. On the other hand the kernel of this operator is identified to the tangent space of $\mathcal{M}((-1,0,0),(1,0,0))$ at ℓ. Since the complex (7.7) is S^1-invariant, we may regard its index as a virtual representation of S^1. Then the index of (7.7) at the element of $\mathcal{M}((-1,0,0),(1,0,0))$ is $\boldsymbol{R} - \boldsymbol{C}$, where $\boldsymbol{R}, \boldsymbol{C}$ are regarded as $S^1 = U(1)$ modules. Then its dimension is negative. But $-(\boldsymbol{R} - \boldsymbol{C})$ is not realized as an actual representation. Hence, as an element of $K_{S^1}(\text{pt})$ (the equivariant K-group of one point), this index is neither "negative" nor "positive". This is the reason that we can not remove this gradient line by S^1-invariant perturbation.

This trouble is essential. So we give up to study equivariant Morse theory in general and consider the following condition for the G invariant Morse function f.

Condition 7.8

(1)   Each critical submanifold R_a is a single G-orbit. We put it G/I_a.

(2)   For each pair of critical submanifolds R_a, R_b, the space $\mathcal{M}(R_a, R_b)$ is a manifold of dimension $\eta(a) - \eta(b) + \dim R_a$.

We remark that Condition 7.8 (1) implies that f is weakly self-indexing.

Example 7.6 above shows that one may not find a G-invariant Morse function satisfying Condition 7.8 by perturbation. From now on we *assume* that the G-invariant function f satisfies Condition 7.8.

Now we want to consider the equivariant cohomology groups : $H_G^k(X; \boldsymbol{Z}) = H^k(X \times_G EG; \boldsymbol{Z})$.

To obtain $H^*(X \times_G EG; \boldsymbol{Z})$ using Morse theory, one trouble is that EG is infinite dimensional. So we take a finite dimensional approximations of it. Namely let EG_v be a sequence of closed finite dimensional manifolds on which G acts freely and that EG_ν is n_ν-connected for some $n_\nu \to \infty$.

For example, if $G = SO(3)$ one can take $EG_i = SO(l+3)/SO(i)$. Then $SO(3) \setminus EG_i$ is a Grassmannian manifold. Hence $EG_\nu = SO(\nu+3)/SO(\nu)$ is a finite dimensional approximation of the total space of the universal bundle $EG \to BG$.

Now we consider the space $X \times_G EG_\nu$. Our (G-invariant) Bott-Morse function f on X induces a Bott-Morse function $\hat{f}_\nu$ on $X \times_G EG_\nu$. Its critical submanifold is given by $R_a \times_G EG_\nu = I_a \setminus EG_\nu$ (by Assumption 7.2 (1)). Our Assumption 7.8 on f implies that $\hat{f}_\nu$ is weakly self-indexing and satisfies 7.8 (2). Hence Theorem 7.5 implies that there exists a spectral sequence $E_{i,j}^{(\nu)k}$ such that

(1) $E_{i,j}^{(\nu)k} \Rightarrow H_{i+j}(X \times_G EG_\nu; \mathbf{Z})$,

(2) $E_{i,j}^{(\nu)1} = \bigoplus_{\eta(a)=i} H_j(I_a \setminus EG_\nu; \mathbf{Z})$.

Now it is easy to take the limit $\nu \to \infty$ of this spectral sequence. Thus we have

Theorem 7.9 :   *Let f be a G-invariant Bott-Morse function on X satisfying Condition (7.8). Then there exists a spectral sequence $E_{i,j}^k$ such that*

(1) $E_{i,j}^k \Rightarrow H_G(X; \mathbf{Z})$,
(2) $E_{i,j}^1 = \bigoplus_{\eta(a)=i} H_j(BI_a; \mathbf{Z})$.

Now we are ready to discuss the case of the Floer homology. As we remarked before, the existence of an equivariant Bott-Morse function satisfying Condition 7.8 is not automatic. But in our situation we can prove the following :

Theorem 7.10   *Let N be an oriented 3-manifold. Then we can perturb the Chern-Simons functional such that it satisfies* Condition 7.8.

This theorem was proved in [Fk 1], though the author did not realize that these properties are essential in equivariant Morse theory at that time.

Thus, using Theorem 7.9 and Theorem 7.10, one can define equivariant Floer theory for oriented 3-manifolds. But in fact there is a trouble to prove that it is independent of the perturbation, in the case when $H_1(N; \mathbf{Q})$ is nonzero. (The trouble is similar to that we explained in Example 7.6.)   But we can overcome this trouble by fixing a basis of $H^1(N;\mathbf{Z}_2)$.

Making use of this kind of equivariant Floer theory, Austin-Braam is now trying to define the relative Donaldson invariant for oriented 3-manifold and to prove the coupling formula.

Higher boundary operator.

We next discuss another point which are to be involved in the theory of Floer homology for general 3-manifolds.

First, we consider the coupling formula for $M = M_1 \times_N M_2$. The coupling formula is to describe the Donaldson invariant of M by those of M_1 and M_2. As we have seen in § 2, the Donaldson invariant of M, a closed manifold, is a polynomial over $H_2(M;\mathbf{Q})$. We recall the Myer-Vietris exact sequence :

$$\to H_2(N;\mathbf{Q}) \to H_2(M_1;\mathbf{Q}) \oplus H_2(M_1;\mathbf{Q}) \to H_2(M;\mathbf{Q}) \to H_1(N;\mathbf{Q}) \to$$

In the case when N is a homology sphere, it implies that $H_2(M;\mathbf{Q}) \cong H_2(M_1;\mathbf{Q}) \oplus H_2(M_2;\mathbf{Q})$. Hence the coupling formula is the formula to describe a relation between $Q_{\ell+\ell'}(\Sigma_1,\cdots,\Sigma_\ell,\Sigma'_1,\cdots,\Sigma'_{\ell'})$, $\Sigma_i \subseteq M_1$, $\Sigma'_i \subseteq M_2$ and $Q_\ell(\Sigma_1,\cdots,\Sigma_\ell) \in HF(N)$ and $Q_{\ell'}(\Sigma'_1,\cdots,\Sigma'_{\ell'}) \in HF(N)$. In the case when, N is not necessary a homology sphere, there is another kind of elements of $H_2(M;\mathbf{Q})$, which come from $H_1(N;\mathbf{Q})$. (See Figure 7.11 below.)

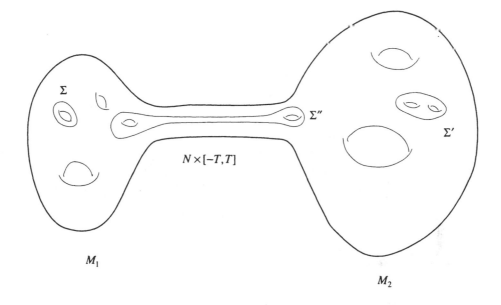

Figure 7.11

Here, $\Sigma'' \cap N \neq \varnothing$ and the homology class of this intersection is the image of $[\Sigma'']$ in $H_1(N;\mathbf{Q})$ by the boundary map of Myer-Vietoris exact sequence.

Hence after cutting the manifold along N, we have elements of $H_2(M_1;\mathbf{Q})$ and those of $H_2(M_1,N;\mathbf{Q})$. For a surface Σ representing an element of $H_2(M_1;\mathbf{Q})$, the bundle $res^{\bullet}_{\Sigma}(L(\Sigma))$ on $\mathcal{B}(M;P)$ can be extended to the compactification of moduli space of ASD connections and a similar discussion as in the proof of Theorem 6.11 works.

However, since $\Sigma'' \cap N \neq \varnothing$, there is a trouble in extending $res^{\bullet}_{\Sigma''}(L(\Sigma''))$ to the compactification of the moduli space of ASD connections.

To isolate this problem from the one related to the singularity, we discuss the case when $a \in R(N)$ is a regular point of $R(N)$. Namely, we assume that $I_a = \{\pm 1\}$ and $H^1(N;su(2)^a) = 0$. Then conclusions of Theorem 6.1 (2) hold in this case. Let $b \in R(N)$ be also a regular point of $R(N)$. Then, by the same argument as Theorem 6.5, we can find that the set $\mathcal{M}(M,P;[b]) \times \overline{\mathcal{M}}([b],[a])$ appears as a part of the end of $\mathcal{M}(M,P;[a])$. So we want to describe the behavior of the bundle $res^{\bullet}_{\Sigma''}(L(\Sigma''))$ there. For simplicity, we assume that $\Sigma'' \cap N$ is a circle γ in N. Then we have

Proposition 7.12 ([Fk1])   *There exists a complex line bundle $\mathcal{L}_{\gamma}(a,b)$ on $\overline{\mathcal{M}}([a],[b])$ for each pair of smooth points of $R(N)$ such that*

(1) $L_\gamma(a,b)$ and $L_\gamma(a,c) \otimes L_\gamma(c,b)$ *can be patched together to give a complex line bundle on the compactification* $C\overline{M}([a],[b])$ *in Corollary 6.7.*

(2)   *Suppose* $\Sigma'' \cap N = \gamma$. *We consider the bundle* $res_{\Sigma''}^*(L(\Sigma''))^{\otimes 2}$ *on* $M(M,P;[a])$. *Then at the end of* $M(M,P;[a])$ *diffeomorphic to* $M(M,P;[b]) \times \overline{M}([b],[a]) \times (0,\infty)$, $L_\gamma(a,b)$ *is isomorphic to* $res_{\Sigma''}^*(L(\Sigma''))^{\otimes 2} \otimes L_\gamma(a,b)$.

We do not explain why $\otimes 2$ appears in (2).

Now we try to handle the chain

$$Q_\ell(\Sigma_1,\cdots,\Sigma_\ell,\Sigma_1',\cdots,\Sigma_{\ell'}')$$

$$= \sum_a \# \left([M(M,P;[a])] \cap \bigcap_{i=1}^{\ell} [res_{\Sigma_i}^{-1}(s_{\Sigma_i}^{-1}(0))] \cap \bigcap_{i=1}^{\ell'} [res_{\Sigma_i'}^{-1}(s_{\Sigma_i'}^{-1}(0))] \right) \cdot [a]$$

as in the last section. We then find that this is not a cycle even if we disregard the trouble coming from the singularity of $\mathcal{B}(N,P)$.

To explain it, let us try to imitate the proof of Theorem 6.11. Then we consider the c a s e   w h e n $\dim M(M,P;[a]) = 2(\ell+\ell')+1$.   T h e n $[M(M,P;[a])] \cap \bigcap_{i=1}^{\ell} [res_{\Sigma_i}^{-1}(s_{\Sigma_i}^{-1}(0))] \cap \bigcap_{i=1}^{\ell'} [res_{\Sigma_i'}^{-1}(s_{\Sigma_i'}^{-1}(0))]$ is a one dimensional manifold. We again disregard the trouble coming from the singular point to study the end of this one dimensional manifold using Proposition 7.12 (2).

Suppose for simplicity $\ell' = 1$. Then there is two kinds of ends namely

$$\left(M(M,P;[c]) \cap \bigcap_{i=1}^{\ell} res_{\Sigma_i}^{-1}(s_{\Sigma_i}^{-1}(0)) \cap res_{\Sigma'}^{-1}(s_{\Sigma'}^{-1}(0)) \right) \times \overline{M}([c],[a])$$

for $\eta(c) = \eta(a) + 1$ and

$$\frac{1}{2} \cdot \left(M(M,P;[b]) \cap \bigcap_{i=1}^{\ell} res_{\Sigma_i}^{-1}(s_{\Sigma_i}^{-1}(0)) \right) \times \left(\overline{M}([b],[a]) \cap s_\gamma^{-1}(0) \right)$$

for $\eta(b) = \eta(a) + 3$. Here s_γ denotes a section of $L_\gamma(b,a)$.

Ends of the first kind are related to the boundary of the chain

$$\langle \partial Q_\ell(\Sigma_1,\cdots,\Sigma_\ell,\Sigma_1',\cdots,\Sigma_{\ell'}'),a \rangle$$

$$= \sum_c \# \left([M(M,P;[c])] \cap \bigcap_{i=1}^{\ell} [res_{\Sigma_i}^{-1}(s_{\Sigma_i}^{-1}(0))] \cap \bigcap_{i=1}^{\ell'} [res_{\Sigma_i'}^{-1}(s_{\Sigma_i'}^{-1}(0))] \right) \cdot \langle \partial c,a \rangle$$

But, since there is a second kind of ends, this boundary may not vanish. Hence we have to consider a variant of the Floer homology which takes into account the line bundle $L_\gamma(b,a)$.

"Definition" 7.13 ([Fk 1])   *Let* $\eta(b) = \eta(a) + 3$. *Then we put*

$$< \partial_\gamma b, a >= \# \overline{\mathcal{M}}([b],[a]) \cap s_\gamma^{-1}(0).$$

In fact, the left hand side does depend on the choice of the section s_γ since $\overline{\mathcal{M}}([b],[a])$ has a boundary. See [Fk 1] for an argument to justify this definition. Now, in the case when there is no effect of singularity, this map satisfies the following relation.

Proposition 7.14   *Suppose that there is no flat connection* x *such that* $\eta(a) > \eta(x) > \eta(b)$ *and* $I_x \neq \{\pm 1\}$. *Then we have*

$$\partial \partial_\gamma + \partial_\gamma \partial = 0.$$

Now we use this Proposition 7.14 to construct the variant of the Floer homology we need. Because of the reason related to the presence of the singularity, we need to assume that $H_1(N; \mathbf{Z})$ is torsion free. Let $k = b_1(N)$. Fix a basis $[\gamma_1], \cdots, [\gamma_k]$ of $H_1(N; \mathbf{Z})$. We introduce a variables $e_1, \cdots, e_k$ of degree -2 and define :

Definition 7.15

$$\begin{cases} CF^{(1)}(N; P) & = & \displaystyle\bigoplus_{\substack{a \in R(N) \\ I_a = \{\pm 1\}}} & \dfrac{\mathbf{Z}[e_1, \cdots, e_k]}{(e_1^2, \cdots, e_k^2)} \cdot [a] \\[2em] \hat{\partial}(1 \otimes [a]) & = & 1 \otimes \partial[a] \\[1em] \hat{\partial}(e_i \otimes [a]) & = & 1 \otimes \partial_{\gamma_i}[a] + e_i \otimes \partial[a] \end{cases}$$

Then Proposition 7.14 implies that $\hat{\partial}\hat{\partial} = 0$. We remark that to carrying out this construction we need to perturb the Chern-Simons functional so that $R(N)$ is discrete. (Hence $R(N)$ denotes the set of critical points of the perturbed Chern-Simons functional.)

We thus obtained a homology group

$$HF_\bullet^{(1)}(N) = H_\bullet((CF^{(1)}(N),\hat{\partial})).$$

We are now in a position to discuss the relative invariant in this situation. We suppose $\pi_1(M_1) = 1$. Let $\Sigma' \subseteq M_1$ with $[\partial\Sigma'] = [\Sigma' \cap N] = \sum \alpha_i[\gamma_i]$. Suppose $\dim \mathcal{M}(M,P;[c]) = 2(\ell+1)$ for $[c] \in R(N)$ with $\eta(c) = m$. Then, for $\Sigma_i \in M_1$, $i = 1,\cdots,\ell$ with $\Sigma_i \cap N = \varnothing$, we consider the following chain in $CF^{(1)}(N)$.

$$
\begin{aligned}
(7.16) \quad & \sum_{\eta(c)=m} 1 \otimes \left([\mathcal{M}(M,P;[c])] \cap \bigcap_{i=1}^{\ell} [res_{\Sigma_i}^{-1}(s_{\Sigma_i}^{-1}(0))] \cap res_{\Sigma'}^{-1}(s_{\Sigma'}^{-1}(0)) \right) \cdot [c] \\
& + \frac{1}{2} \cdot \sum_i \sum_{\eta(b)=m-2} \alpha_i e_i \otimes \left([\mathcal{M}(M,P;[b])] \cap \bigcap_{i=1}^{\ell} [res_{\Sigma_i}^{-1}(s_{\Sigma_i}^{-1}(0))] \right) \cdot [b]
\end{aligned}
$$

Then using Proposition 7.12, we can mimic the proof of Theorem 6.11 to prove that (7.16) is a cycle. We denote this cycle by $Q_\ell(\Sigma_1,\cdots,\Sigma_\ell,\Sigma') \in HF_\bullet^{(1)}(N)$.

In this manner one may describe the relative invariant and may prove the coupling formula in certain cases. But, in fact, there arises a troubles from the singularity of $\mathcal{B}(N,P)$. As a consequence, this technique, with only minor modification, works only in the case when $H_1(N;\mathbf{Z})$ is torsion free and there are at most three homology classes intersecting with N. So, to get a better picture, one need to combine this idea with that of the equivariant Floer homology.

However, if we consider a nontrivial $SO(3)$-bundle over N in place of the trivial $SU(2)$-bundle, there are cases where no singularity appear on $\mathcal{B}(N,P)$. In this case, P.Braam and S.Donaldson [BD], uses the construction discussed above to define the relative invariant and prove coupling formula.

Degeneration at infinity of ASD equation.

There are two kinds of approaches to study the relative Donaldson invariant for 4-manifolds with boundary. In the approach we have been discussing so far, we perturbed the ASD equation (or equivalently the Chern-Simons invariant) at the end in order to make the degeneration of the ASD equation as mild as possible. In this way, we hope to get the general picture and obtain necessary machinery to have a good understanding of the phenomenon.

However there is another approach where one does not perturb equation.

In fact, even if one is interested only in the equation after perturbation, one needs sometimes the result before perturbation to obtain explicit results. Let me mention one example of it.

Let us take, for example, $N = T^3$. It has no flat connection for which $I_a = \{\pm 1\}$. So, if we try to prove the coupling formula without much care, then, by the constructions we have been discussing, it seems that it would imply that under some assumptions the Donaldson invariant of a 4-manifold $M = M_1 \cup_{T^3} M_2$ vanishes, even for the class which intersect T^3. (Or better to say it would vanish if there are at most 3 classes which intersects T^3.) However as Morgan and Mrowka showed that this is not true. As Mrowka pointed out to the author (in 1992 September at Oberwolfach), the reason is as follows. We perturb the Chern-Simons functional to $cs + \varepsilon \sum_{i=1}^{3} \mathrm{Tr}(h_{\gamma_i})$, (where $\gamma_i, i = 1, 2, 3$ is a basis of $H_1(T^3; \mathbf{Z})$.) Then after perturbation, we have 8 critical points for which $I_a = \{\pm 1\}$. (They corresponds to elements of $H^1(T^3; \mathbf{Z}_2)$.) But, using Mrowka's result, one can prove that there will be a new critical point for which $I_a \neq \{\pm 1\}$. Hence, to see what happens after perturbation and calculate the Floer homology, we need to see carefully the behavior of the Chern-Simons functional before perturbation.

Now we discuss here a few results obtained about it. But, because of the ability of the author, we discuss it only quite briefly.

Let us go back to the situation of finite dimensional Morse theory. We consider a function $f : X \to \mathbf{R}$ on a finite dimensional manifold X. The most nondegenerate case is that f is a Morse function, which corresponds the case we discussed in § 6.

The next nondegenerate case is when f is a Bott-Morse function. Let us discuss this case. The finite dimensional version of it is already discussed in this section. Lemma 7.4 is the result corresponding to Lemma 4.5. We next generalize Lemma 4.7 to our Morse Bott situation. We use the same terminology as Lemma 7.4.

We first recall the following diagram

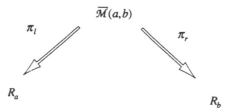

Here π_l and π_r are the maps $a \mapsto a(\pm\infty)$. To avoid confusion we write them as

$\pi_{\ell,(a,b)}$ and $\pi_{r,(a,b)}$.

Assumption 7.17

(1) $\pi_{\ell,(a,b)}$ is transversal to $\pi_{r,(c,a)}$ for each a,b,c.

(2) f is weakly self-indexing.

Lemma 7.18   *If* Assumption 7.17 *is satisfied, then we can compactify the moduli space* $\overline{\mathcal{M}}(a,b)$ *such that*

$$\partial\overline{\mathcal{M}}(a,b) = \bigcup_c \overline{\mathcal{M}}(a,c) \times_{R_c} \overline{\mathcal{M}}(c,b).$$

In fact, the proof of Theorem 7.5 is based on Lemma 7.18.

In finite dimensional case the proof proceeds rather straightforward. Now we discuss the infinite dimensional version of it.

We consider the 3-manifold N and, for simplicity, we assume that it is a homology 3-sphere. Then the condition that cs is Bott-Morse function is translated as follows.

Assumption 7.19

(1) $R(N)$ is a smooth manifold.

(2)   For each $[a] \in R(N)$, we have $H^1(N; su(2)^a) \cong T_{[a]}R(N)$.

Then one can prove an analogy of Theorem 6.1. We assume that N satisfy Assumption 7.19. Let M be a 4-manifold which bounds N. We put it a complete Riemannian metric on M as in § 6. We consider

$$\mathcal{M}(M,P;R_a) \quad = \quad \left\{ ([A]) \left| \begin{array}{l} [A] \in \mathcal{B}(M,P), \\ -F_A = *F_A, \\ A|_{\{\infty\} \times N} \in R_a. \end{array} \right. \right\}$$

and

$$\mathcal{M}(M,P) \quad = \quad \left\{ ([A]) \left| \begin{array}{l} [A] \in \mathcal{B}(M,P), \\ -F_A = *F_A, \\ \int_M \|F_A\|^2 < \infty. \end{array} \right. \right\}$$

Then we have :

Theorem 7.20 (Mrowka [M1] see also [Fk?])

(1) $\mathcal{M}(M,P) = \bigcup_a \mathcal{M}(M,P;R_a).$

(2) $\mathcal{M}(M,P;R_a)$ *is a smooth manifold for generic metric on* M.

(3)   *If* $A \in \mathcal{M}(M,P;R_a)$, *then there exists uniquely* $[x] \in R_a$ *such that*

$$\|A - x\| \le Ce^{-ct}.$$

(4)   *The map* $A \mapsto a$ *is a smooth map.*

Let us denote by $\pi_{r,M,a}$ the map (4). In the case when $M = N \times \mathbf{R}$ we will have an analogy. But, in general, we can not perturb the metric of N so that (2) holds. Hence (as in the case when cs is a Morse function), we have to take artificial perturbation to keep the $\mathbf{R}$-invariance. But we can do it outside the set $R(N)$. So, for simplicity, we assume that this is done already. We put

$$\mathcal{M}(a,b) = \left\{ \ell : \mathbf{R} \to \mathcal{B}(N) \left| \begin{array}{l} \ell(-\infty) \in R_a, \\ \ell(\infty) \in R_b, \\ \dfrac{d\ell}{dt} = -\text{grad } cs \end{array} \right. \right\}$$

We have :

(1) $\mathcal{M}(a,b)$ is a manifold of dimension $\eta(a) - \eta(b)$.

(2) $\pi_{\ell,(a,b)}:\mathcal{M}(a,b) \to R_a$, $\ell \mapsto \ell(-\infty)$ and $\pi_{r,(a,b)}:\mathcal{M}(a,b) \to R_b$, $\ell \mapsto \ell(+\infty)$ are smooth maps.

Now we consider an analogy of Theorem 6.5. We assume Assumption 7.19 and that $\pi_{r,M,a}$ is transversal to $\pi_{l,(a,b)}$ for each a,b

Theorem 7.21 (Mrowka [Mr] see also [Fk2])   *There exists a space* $CM(M,P;R_a)$ *such that*

(1) $\mathcal{M}(M,P;R_a)$ *is dense in* $CM(M,P;R_a)$,

(2)   *the complement of the union of*

$$\bigcup_b \hat{\mathcal{M}}(M,P;R_b) \times_{R_b} \overline{\mathcal{M}}([b],[a])$$

and an $SO(3)$-bundle over $\mathcal{M}(M,P;R_0) \times \overline{\mathcal{M}}([0],[a])$ in $C\mathcal{M}(M,P;R_a) - \mathcal{M}(M,P;a)$ is a space of $\mathrm{codim} \geq 2$ in $C\mathcal{M}(M,P;R_a)$.

(3)   If $[A_i] \in \mathcal{M}(M,P;R_a)$ is a sequence such that $\sup|F_{A_i}|$ is bounded uniformly on i, then $[A_i]$ has a subsequence converging in $C\mathcal{M}(M,P;R_a)$.

Thus we described the case when the Chern-Simons functional is a Bott-Morse function.

Then what happens when there arises more serious degeneration ? Morgan-Mrowka-Rubermann [MMR] uses the idea of center manifold for this purpose.

Let us discuss it (only in finite dimensional case.)

Let $f : X \to \boldsymbol{R}$ be a function and suppose $p \in X$ with $df(p) = 0$. We consider $\mathcal{H} = \left\{ U \in T_p(X) \mid Hess_f(U) = 0 \right\}$. Since we do not assume that f is Bott-Morse function, $\mathcal{H}$ may not coincide with the tangent bundle of $Cr(f)$ at p. (Furthermore, $Cr(f)$ may not have a tangent space at p, since it may be singular.)

In this situation one can prove :

Theorem 7.22 (Center manifold theorem)   *There exists an (open) submanifold H of X such that*

(1) $T_p H = \mathcal{H}$.

(2) H is $-\mathrm{grad}\, f$ invariant.

(3)   *For each orbit ℓ in X of $-\mathrm{grad}\, f$ such that $\lim_{t \to \infty} \ell(t) = p$, there is an orbit ℓ' of $-\mathrm{grad}\, f$ in H such that*

(7.23) $$d(\ell(t), \ell'(t)) \leq Ce^{-ct}$$

holds for some positive constants C, c.

As already discussed, the Formula (7.23) holds for $\ell' \equiv p$ in the case when f is Bott-Morse function. So, roughly speaking, Theorem 7.22 says that the effect of the degeneration of the function f can be understood by considering the gradient vector field on H. In finite dimensional case, this theorem is rather standard. See for example [LA], for its proof.

Morgan-Mrowka-Rubermann [MMR] proved the corresponding result in the situation where f is the Chern-Simons functional. In fact the result is especially interesting in the infinite dimensional situation. In fact the manifold H is of finite dimension, since $\mathcal{H}$ is finite dimension by the ellipticity of the equation. Therefore we can reduce the problem to finite dimensional situation using that infinite dimensional analogue of Theorem 7.22.

Before those works related to center manifolds and gauge theory had been done, Taubes studied the case when N is an S^1-bundle over a surface in detail. Using it, Taubes claimed that he constructed the relative invariant and proved the coupling formula in that case, recently.

Thus we have discussed some of the building blocks of the general Floer theory for general 3-manifolds. I think that the story about them is now very close to be completed and in a few years and we will know the precise picture of what are the 3+4 dimensional topological field theory based on the Yang-Mills equation.

References

[A]   M.Atiyah, *New invariants for 3 and 4 dimensional manifolds* , In "Mathematical heritage of Hermann Weyl", Proceeding of Symposia in Pure Mathematics **48**, 285 - 299.

[AHD]   M.Atiyah,N.Hitchin,I.Singer, *Self duality in four dimensional Riemannian geometry*, Porceedings of the Royal Society of London, Series A, **362** (1978), 425 - 461.

[AM]   S.Akbult, J.Macarthy, "Casson's invariant for oriented homology 3-spheres", Princeton University Press, Princeton, 1990.

[AB1]   D.Austin,P.Braam, *More-Bott theory and equivariant cohomology*, preprint.

[AB2]   D.Austin,P.Braam, *Equivariant (co)homology and gluing Donaldson polynomials*, preprint.

[BD]   P.Braam,S.Donaldson, *Fukaya-Floer homology and gluing formulae for polynomial invariant*, preprint.

[D1]   S.Donaldson, *An application of gauge theory to the topology of 4-manifolds*, J. Differential Geom. **18** (1983), 269 - 316.

[D2]   _____, *The orientation of Yang-Mills moduli spaces and 4-manifold topology*, J. Differential Geom. **26** (1986), 397 - 428.

[D3]   _____, *Connection, Cohomology, and intersection forms of 4-manifolds*, J. Differential Geom. **24** (1986), 275 - 341.

[D4]   _____, *Polynomal invariant for smooth 4-manifold*, Topology **29** (1990), 141 - 168.

[D5]   _____, *Irrationality and h-cobordism conjecture* , J. Differential Geom. **26** (1987), 141 - 168.

[DK]   S.Donaldson,P.Kronheimer, *"The Geometry of four manifolds"*, Oxford Universitiy Press, 1990.

[DFK]   S.Donaldson,M.Furuta,D.Kotschick, *Floer homology groups in Yang-Mills Theory*, in preparation.

[F1]   A.Floer, *An instanton invariant for three manifold* , Comm. Math. Phys. **118** (1988),215 - 240.

[F2]   _____, *Symplectic fixed points and holomorphic spheres*, Comm. Math. Phys. **120** (1989) 575 - 611.

[FU]   D.Freed, K.Uhlenbeck, "Instantons and Four manifolds", Springer, New York, 1984.

[Fk1]   K.Fukaya, *Floer homology for oriented three manifolds*, in "Advanced studies in pure mathematics, 20 "Aspects of Low dimensional Manifolds" ed. by Matsumoto, Morita, 1992.

[Fk2]   K.Fukaya, *Floer homology of the connected sum of homology three spheres*, preprint.

[FFO]   K.Fukaya,M.Furuta,H.Ohta, *Some variants of instanton cohomology groups and intersection form of 4-manifolds with boundary*, in preparation.

[Fr]   K.Furuta, *An analogue of Floer homology for lens spaces*, preprint.

[K]   D.Kotchik, *On manifold homeomorphic to* $CP^2 \# 8\overline{CP}^2$, Invent. Math. **95** (1989),591 - 600.

[Li]   W.Li, *Connected sums, Floer homology, Spectral sequence*, preprint,

[LA]   G.Iooss,M.Adelmyer, *" Topics in Bifurcation theory"*, World Scientific, Singapore, 1992.

[M]   T.Mrowka, *A local Myer-Vietris Principle for Yang-Mills moduli spaces*, Thesis, Harvard Univ. (1989).

[MMR] J.Morgan,T.Mrowka,D.Ruberman, *The* L^2-*moduli space and a VanishingTheorem for Donaldson invariants* , preprint.

[S]   A.Schwartz, On regular solution of Euclidean invariants and Yang-Mills equations, Phy Letter 6713 (1977), 172 - 174.

[T1]   C.Taubes, *Self-dual connections on non-selfdual manifolds*, J. Differential Geom. **17** (1982), 139 - 170.

[T2]   _____, *Gauge theory on asymptotically periodic 4-manifolds*, J. Differential Geom. **23** (1987), 364 - 430.

[T3]   _____, *Casson's invariant and gauge theory*, J. Differential Geom. **31** (1989), 547 - 599.

[T4]   _____, *L²-moduli spaces on 4-manifolds with cylindrical ends* I, preprint.

[T5]   _____, *A symplicial model for Donaldson Floer thory*, preprint.

[T6]   _____, *The role of reducibles in Donaldson-Floer theory*, preprint.

[U1]   K.Uhlenbeck, *Removable singularities in Yang-Mills fields*, Comm. Math. Phys., **83** (1988), 11 - 29.

[U2]   _____, *Connection with* L^p-*bounds on curvature*, Comm. Math. Phys.,. **83** (1988), 31 - 42.

[Wa]   G.Wasserman, *Equivariant differential topology*, Topology **8** (1969), 127 - 150.

[W1]   E.Witten, *Supersymmetry and Morse theory*, J. Differential Geom. **17** (1982), 661 - 692.

[W2]   E.Witten, *Topological quantum field theory*, Comm. Math. Phys.,. **117** (1988), 353 - 386.

[Y]   T.Yoshida, *Floer homology and splitting of manifolds*, Ann. of Math,. **134** (1991), 277 - 324.

Theorems on the Regularity and Singularity of Minimal Surfaces and Harmonic Maps

Leon Simon[*]

Department of Mathematics
Stanford University
Stanford, CA 94305
USA

Introduction

These lectures are meant as an introduction to the analytic aspects of the study of regularity properties and singularities of minimal surfaces and harmonic maps.

Because it is technically simpler to do so, we concentrate mainly on harmonic maps, but the reader should keep in mind that all the techniques described here for harmonic maps have very close analogues in the theory of minimal surfaces. We do in the final lecture at least state the main results on singularities of minimal surfaces.

The lectures will cover the following main topics:

Lecture 1: Basic Definitions, and the ϵ-Regularity and Compactness Theorems.
Lecture 2: Tangent Maps and Approximation Properties of Subsets of $\mathbf{R}^n$
Lecture 3: Asymptotics near Singular Points.
Lecture 4: Recent Results on Smoothness and Rectifiability of the Singular Set.

LECTURE 1

1. Basic Definitions and the ϵ-Regularity and Compactness Theorems

Assume that Ω is an open subset of $\mathbf{R}^n$, $n \geq 2$. For convenience we assume that Ω has the standard Euclidean metric; only minor technical modifications are needed to extend all the main results here to the case when Ω is equipped with an arbitrary smooth Riemannian metric. N ("the target") will denote a

[*] Partially supported by NSF grant DMS-9207704 at Stanford University; part of the work described in Lectures 3, 4 of these notes was carried out during visits to the Pure Mathematics Department, University of Adelaide, and ETH, Zürich. The author is grateful for the hospitality of these institutions, and also to the organizing committee for the conference on Geometry and Global Analysis held at Tohoku University, Sendai, Japan.

smooth compact Riemannian manifold of dimension $p \geq 2$ which is isometrically embedded (without loss of generality by the Nash embedding theorem) in some Euclidean space $\mathbf{R}^P$. We look at maps u of Ω into N; such a map will always be thought of as a map $u = (u^1, \ldots, u^P) : \Omega \to \mathbf{R}^P$ with the additional property that $u(\Omega) \subset N$. We do not assume that u is smooth—in fact we make only the minimal assumption necessary to ensure that the energy of u is well-defined. Thus we assume only that $u \in W^{1,2}_{\mathrm{loc}}(\Omega; N)$, the subset of the Sobolev space $W^{1,2}_{\mathrm{loc}}(\Omega; \mathbf{R}^P)$ consisting of all functions $u = (u^1, \ldots, u^P) \in W^{1,2}_{\mathrm{loc}}(\Omega; \mathbf{R}^P)$ such that $u(x) \in N$ a.e. $x \in \Omega$. (Recall that $u \in W^{1,2}_{\mathrm{loc}}(\Omega; \mathbf{R}^P)$ means that $u \in L^2_{\mathrm{loc}}(\Omega; \mathbf{R}^P)$ and that the gradient $(D_i u^j)_{i=1,\ldots,n,\, j=1,\ldots,P} \in L^2_{\mathrm{loc}}(\Omega; \mathbf{R}^{nP})$). Then the energy $\mathcal{E}_{B_\rho(y)}(u)$ of u in a ball $B_\rho(y) \equiv \{x : |x - y| < \rho\}$ with $\overline{B}_\rho(y) \subset \Omega$ is defined by

$$\mathcal{E}_{B_\rho(y)}(u) = \int_{B_\rho(y)} |Du|^2,$$

where $|Du|^2 = \sum_{i=1}^n \sum_{j=1}^P (D_i u^j)^2$. We study maps which minimize energy in Ω in the sense that, for each open $\widetilde{\Omega} \subset\subset \Omega$ (i.e. closure of $\widetilde{\Omega}$ is a compact subset of Ω),

$$\mathcal{E}_{\widetilde{\Omega}}(u) \leq \mathcal{E}_{\widetilde{\Omega}}(w),$$

for every $w \in W^{1,2}_{\mathrm{loc}}(\Omega; N)$ with $w \equiv u$ in $\Omega \backslash \widetilde{\Omega}$. Such u will be called an energy minimizing map into N.

Now suppose $u \in W^{1,2}_{\mathrm{loc}}(\Omega; N)$ is energy minimizing, $\overline{B}_\rho(y) \subset \Omega$, and suppose that for some $\delta > 0$ we have a 1-parameter family $\{u_s\}_{s \in (-\delta, \delta)}$ of maps of $B_\rho(y)$ into N such that $Du_s \in L^2(\Omega)$ and $u_s \equiv u$ in a neighbourhood of $\partial B_\rho(y)$ for each $s \in (-\delta, \delta)$, and $u_0 = u$. Then by definition of minimizing we have $\mathcal{E}_{B_\rho(y)}(u_s)$ takes a minimum at $s = 0$, and hence

$$* \qquad \qquad \frac{d\mathcal{E}(u_s)}{ds}\Big|_{s=0} = 0$$

whenever the derivative on the left exists. The derivative on the left is called the first variation of $\mathcal{E}_{B_\rho(y)}$ relative to the given family; the family $\{u_s\}$ itself is called an (admissible) variation of u. There are two important kinds of variations of u as follows:

(i): Variations of the form

$$** \qquad \qquad u_s = \Pi \circ (u + s\zeta),$$

where $\zeta = (\zeta^1, \ldots, \zeta^P)$ with each $\zeta^j \in C_c^\infty(B_\rho(y))$ and where Π is the nearest point projection onto N. (Notice that nearest point projection onto N is well-defined and smooth in some open W containing N, and hence u_s defined in (i) is an admissible variation for $|s|$ small enough). Now by direct computation (using the fact that $d\Pi_z$ is the othogonal projection of $\mathbf{R}^p$ onto $T_z N$ and that second fundamental form A_z of N at a point $z \in N$ is related to the Hessian of Π according to the formulae

$$\text{Hess}_z\, \Pi(\eta, \tau) = -\eta \cdot A_z(\tau, \cdot), \quad \text{Hess}_z\, \Pi(\tau_1, \tau_2) = -A_z(\tau_1, \tau_2)$$

for any $\tau, \tau_1, \tau_2 \in T_z N$ and any $\eta \in (T_z N)^{\perp}$, we easily check that $D_i u_s = D_i u + s((D_i \zeta)^T - \zeta \cdot A_u(D_i u, \cdot) - A_u(D_i u, \zeta^T)) + R$, where $|R| \leq Cs^2(1 + |Du|^2)$, and hence for such a variation $*$ implies the integral identity

$$(1) \qquad \int_\Omega \sum_{i=1}^n \left(D_i u \cdot D_i \zeta - \zeta \cdot A_u(D_i u, D_i u)\right) = 0$$

for any ζ as above. Notice that if u is C^2 we can integrate by parts here and use the fact that ζ is an arbitrary C^∞ function in order to deduce the equation

$$(1)' \qquad \Delta u + \sum_{i=1}^n A_u(D_i u, D_i u) = 0,$$

where Δu means simply $(\Delta u^1, \ldots, \Delta u^P)$. The identity (1) is called the weak form of the equation $(1)'$; of course if u is not C^2 the equation $(1)'$ makes no sense classically, and <u>must</u> be interpreted in the weak sense (1). Notice that in case $u \in C^2$, (1) says simply

$$(\Delta u)^T = 0$$

at a given point $x \in B_\rho(y)$, where $(\Delta u)^T$ means orthogonal projection of $\Delta u(x)$ onto the tangent space $T_{u(x)} N$ of N at the image point $u(x)$. We shall make no specific use of this in the sequel.

(ii): Variations of the form

$$u_s(x) = u(x + s\zeta(x)),$$

where $\zeta = (\zeta^1, \ldots, \zeta^n)$ with each $\zeta^j \in C_c^\infty(B_\rho(y))$. Then $D_i u_s(x) = \sum_{j=1}^n D_i u(x + s\zeta) + s D_i \zeta^j D_j u(x + s\zeta)$, and hence after making the change of variable $\xi = x + s\zeta$ (which gives a C^∞ diffeomorphism of $B_\rho(y)$ onto itself in case $|s|$ is small enough) in this case $*$ implies

$$(2) \qquad \int_{B_\rho(y)} \sum_{i,j=1}^n \left(|Du|^2 \delta_{ij} - 2D_i u \cdot D_j u\right) D_i \zeta^j = 0.$$

The identities (1), (2) are of central importance in the study of energy minimizing maps. Notice that if $u \in C^2$ we can integrate by parts in (1) in order to deduce that (1) <u>implies</u> (2); it is however false that (1) implies (2) in case Du is merely in L^2 (and there are simple examples to illustrate this). One calls a map u into N which satisfies (1) a "weakly harmonic map", while a map which satisfies both (1) and (2) is usually referred to as a "stationary

harmonic map". The above discussion thus proves that energy minimizing implies stationary harmonic. We shall not here discuss weakly harmonic maps or
stationary harmonic maps, but we do want to mention that weakly harmonic
maps admit far worse singularities than the energy minimizing maps (see e.g.
[HL1,2]), except in the case $n = 2$ when there are no singularities at all—we
show this below in the case of minimizing maps, and refer to recent work of
F. Hélein [HF] for the general case of weakly harmonic maps. For stationary
harmonic maps it is known that the singular set is codimension at least 2
(due to Evans [E] in the case when the target is a sphere and for general
target due to Bethuel [B]), but so far no result about the actual structure of
the singular set has been obtained in this case.

An important consequence of the variational identity (2) is the "monotonicity identity"

$$(3) \qquad \rho^{2-n} \int_{B_\rho(y)} |Du|^2 - \sigma^{2-n} \int_{B_\sigma(y)} |Du|^2 = 2 \int_{B_\rho(y) \backslash B_\sigma(y)} r^{2-n} \left|\frac{\partial u}{\partial r}\right|^2,$$

valid for any $0 < \sigma < \rho < \rho_0$, provided $\overline{B}_{\rho_0}(y) \subset \Omega$, where $r = |x - y|$ and
$\partial/\partial r$ means directional derivative in the radial direction $|x - y|^{-1}(x - y)$.
Since it is a key tool in the study of energy minimizing maps, we give the
proof of this identity.

Proof: First recall a general fact from analysis—Viz. if a_j are L^1 functions on
$B_{\rho_0}(y)$ and if $\int_{B_{\rho_0}(y)} \sum_{j=1}^n a^j D_j \zeta = 0$ for each ζ which is C^∞ with compact
support in $B_{\rho_0}(y)$, then, for almost all $\rho \in (0, \rho_0)$, $\int_{B_\rho(y)} \sum_{j=1}^n a_j D_j \zeta =$
$\int_{\partial B_\rho(y)} \eta \cdot a\zeta$ for any $\zeta \in C^\infty(\overline{B}_\rho(y))$, where $a = (a^1, \ldots, a^n)$ and $\eta (\equiv$
$\rho^{-1}(x - y))$ is the outward pointing unit normal of $\partial B_\rho(y)$. (This fact is
easily checked by approximating the characteristic function of the ball $B_\rho(y)$
by C^∞ functions with compact support.) Using this in the identity (2), we
obtain (for almost all $\rho \in (0, \rho_0)$) that

$$\int_{B_\rho(y)} \sum_{i,j=1}^n (|Du|^2 \delta_{ij} - 2D_i u \cdot D_j u) D_i \zeta^j =$$

$$\int_{\partial B_\rho(y)} \sum_{i,j=1}^n ((|Du|^2 \delta_{ij} - 2D_i u \cdot D_j u) \rho^{-1}(x^i - y^i) \zeta^j).$$

In this identity we choose $\zeta^j(x) \equiv x^j - y^j$, so $D_i \zeta^j = \delta_{ij}$ and we obtain

$$(n - 2) \int_{B_\rho(y)} |Du|^2 = \rho \int_{\partial B_\rho(y)} (|Du|^2 - 2|\partial u/\partial r|^2).$$

Now by multiplying through by the factor ρ^{1-n} and noting that $\int_{\partial B_\rho} f =$
$\frac{d}{d\rho} \int_{B_\rho} f$ for almost all ρ, we obtain the differential identity

$$\frac{d}{d\rho}\left(\rho^{2-n}\int_{B_\rho(y)}|Du|^2\right) = 2\int_{\partial B_\rho(y)} r^{2-n}\left|\frac{\partial u}{\partial r}\right|^2$$

for almost all $\rho \in (0,\rho_0)$. Since $\int_{B_\rho} f$ is an absolutely continuous function of ρ and since $\int_\sigma^\rho \int_{\partial B_r} f = \int_{B_\rho \setminus B_\sigma} f$ (for any L^1-function f), we can now integrate to give the required monotonicity identity.

Notice that since the right side of ‡ is non-negative, we have in particular that

(4) $\qquad \rho^{2-n}\int_{B_\rho(y)}|Du|^2$ is an increasing function of ρ for $\rho \in (0,\rho_0)$,

and hence that the limit as $\rho \to 0$ of $\rho^{2-n}\int_{B_\rho(y)}|Du|^2$ exists. We define the density function Θ_u of u on Ω to be this limit; thus

(5) $$\Theta_u(y) = \lim_{\rho\downarrow 0}\rho^{2-n}\int_{B_\rho(y)}|Du|^2.$$

Finally note that since the left side of (3) has a limit as $\sigma \downarrow 0$ we have $\int_{B_\rho(y)} r^{2-n}|\partial u/\partial r|^2 < \infty$ and

(6) $$\rho^{2-n}\int_{B_\rho(y)}|Du|^2 - \Theta_u(y) = 2\int_{B_\rho(y)} r^{2-n}\left|\frac{\partial u}{\partial r}\right|^2.$$

2. The ϵ-Regularity and Compactness Theorems

The key theorems in a first study of regularity and singularity of energy minimizing maps are the ϵ-regularity theorem and the compactness theorem. The ϵ-regularity theorem is due to Schoen-Uhlenbeck [SU] (the same result for maps into a single coordinate chart of the target manifold was established at about the same time by Giaquinta-Giusti [GG]), and is completely analogous to the ϵ-regularity theorems proved for various classes of minimal surfaces by De Giorgi [DeG], Reifenberg [R], Almgren [A2], Allard [AW], and Schoen-Simon [SS].

Theorem 1 (ϵ-regularity theorem). *If Λ is any given fixed positive constant, then there is $\epsilon > 0$, depending only on n, N, Λ, such that if u is an energy minimizing map into N if $\overline{B}_\rho(y) \subset \Omega$, if $\rho^{2-n}\int_{B_\rho(y)}|Du|^2 \le \Lambda$, and if*

$$\min_{\lambda\in\mathbf{R}^P}\rho^{-n}\int_{B_\rho(y)}|u-\lambda|^2 < \epsilon,$$

then $y \in \mathrm{reg}\, u$, and furthermore for each $j \ge 0$

$$\sup_{B_{\rho/2}(y)} \rho^j|D^j u| \le C,$$

where C depends only on n, N, j.

Remark. Given the bound $\rho^{2-n} \int_{B_\rho(y)} |Du|^2 \leq \Lambda$ on the scaled energy for some given ball $B_\rho(y)$ with closure contained in Ω, the theorem says roughly that if the mean square deviation of u away from some constant vector is sufficiently small in this ball, then y is a regular point of u and all the derivatives of u are controlled in the ball of radius $\rho/2$ and center y.

Next we have a compactness theorem due to Luckhaus [Lu1], [Lu2] extending partial results of [SU] and [HL2].

Theorem 2 (Compactness theorem.). *If $\{u_j\}$ is a sequence of energy minimizing maps from Ω into N with $\sup_j \int_{B_\rho(y)} |Du_j|^2 < \infty$ for each ball $B_\rho(y)$ with $\overline{B}_\rho(y) \subset \Omega$, then there is a subsequence $\{u_{j'}\}$ and a minimizing harmonic map u of Ω into N such that $u_{j'}$, $Du_{j'}$ converge in L^2 locally on Ω to u, Du respectively.*

Note: In particular the energy $\int_{B_\rho(y)} |Du_{j'}|^2$ converges to $\int_{B_\rho(y)} |Du|^2$ for each ball $\overline{B}_\rho(y) \subset \Omega$.

Remarks on the proofs of Theorems 1, 2: Notice that, by the general Rellich compactness theorem for Sobolev space there is a $W^{1,2}_{loc}(\Omega; \mathbf{R}^p)$ function u such that $u_{j'}$ converges in L^2 to u on compact subsets of Ω and $Du_{j'}$ converges locally weakly in L^2 to Du in Ω. Of course then u maps into N (in the sense that $u(x) \in N$ a.e. $x \in \Omega$) because a subsequence of the subsequence $u_{j'}$ converges pointwise a.e. to u, so that $u \in W^{1,2}_{loc}(\Omega; N)$. Thus the main content of Theorem 2 is that $Du_{j'}$ converges in L^2 and that u is minimizing. The main difficulty in proving these latter facts is that on a given ball $B_\rho(y)$ with closure contained in Ω, the values of u_j, u will differ slightly near the boundary $\partial B_\rho(y)$, so that we are not able to use the definition of energy minimizing directly.

However there is a lemma of Luckhaus [Lu1,2] which makes it possible to compare the energies of functions with different boundary values in a manner which is sufficiently precise to imply the above compactness result. We shall not state Luckhaus' lemma here, but we do note that one of its consequences is that there exists $\delta_0 = \delta_0(n, N) > 0$ such that if w_1, $w_2 \in W^{1,2}(B_\rho(y); N)$ with $\mathcal{E}_{B_\rho(y)}(w_j) \leq \Lambda$, $j = 1, 2$, if $\epsilon > 0$, and if $\epsilon^{-2n} \int_{B_\rho(y) \setminus B_{(1-\epsilon)\rho}(y)} |w_1 - w_2|^2 < \delta_0^2$, then there is $w \in W^{1,2}(B_\rho(y) \setminus B_{(1-\epsilon)\rho}(y); N)$ such that $w = w_1$ in a neighbourhood of $\partial B_{(1-\epsilon)\rho}(y)$, $w = w_2$ in a neighbourhood of $\partial B_\rho(y)$, and

$$\int_{B_{(1+\epsilon)\rho}(y) \setminus B_\rho(y)} |Dw|^2 \leq C \int_{B_\rho(y) \setminus B_{(1-\epsilon)\rho}} (|Dw_1|^2 + |Dw_2|^2)$$
$$+ C\epsilon^{-2}\rho^{-2} \int_{B_\rho(y) \setminus B_{(1-\epsilon)\rho}} |w_1 - w_2|^2.$$

Using this lemma with u_k, u_ℓ in place of w_1, w_2 respectivley, it is then elementary to prove the compactness theorem by directly using the definition of energy minimizing. For the complete details we refer to the lecture notes [SL4], [SL5].

Next we outline the proof of the ϵ-regularity theorem. The method is based on "blowing up" (i.e. harmonic approximation), a method going back to De Giorgi. The precise method we use here follows the lecture notes [SL5], which simplifies previous proofs somewhat.

An initial (and non-trivial) step in the proof is to establish the inequality

$$(1) \qquad \rho^{2-n} \int_{B_{\rho/2}(y)} |Du|^2 \leq C\rho^{-n} \int_{B_\rho(y)} |u - \lambda_{y,\rho}|^2.$$

with C depending only on n, N, Λ, where Λ is any upper bound for the scaled energy $\rho^{2-n} \int_{B_\rho(y)} |Du|^2$. This is a kind of "reverse Poincaré" inequality for the energy minimizer u. The proof uses the same Luckhaus lemma mentioned in the above proof of the compactness theorem together with the monotonicity result §1(4) above and the definition of energy minimizing. We shall not prove (1) here, since the complete details can be found in the lecture notes [SL5].

It is an interesting point that the inequality (1) is one of the few places where the theory of minimal surfaces and the theory of harmonic maps are not perfectly analogous, because there is a very easily proved analogue of this reverse Poincaré inequality[1] for general minimal surfaces (or even for general stationary varifolds), but no such inequality for stationary harmonic maps (without the energy minimizing hypothesis assumed here) is known.

To continue the proof of Theorem 1, we write down the weak form of the Euler-Lagrange equation, as described in §1(1) above:

$$\sum_{j=1}^n \int_\Omega D_j u \cdot D_j \zeta = \int_\Omega \zeta \cdot \sum_{j=1}^n A_u(D_j u, D_j u),$$

valid for any $\zeta = (\zeta^1, \ldots, \zeta^P) \in C_c^\infty(\Omega; \mathbf{R}^P)$. Now $\sup_{z \in N} |A_z(\tau, \tau)| \leq C|\tau|^2$, with C a constant depending only on N, and hence if support of $\zeta \subset B_{\rho/2}$ this gives

$$\left| \int_{B_\rho(y)} Du \cdot D\zeta \right| \leq C \sup |\zeta| \int_{B_{\rho/2}(y)} |Du|^2.$$

By virtue of the reverse Poincaré inequality (1) this implies

$$\left| \int_{B_\rho(y)} Du \cdot D\zeta \right| \leq C\rho^{-2} \sup |\zeta| \int_{B_\rho(y)} |u - \lambda_{y,\rho}|^2.$$

With $v = C^{-1}(\rho^{-n} \int_{B_\rho(y)} |u - \lambda_{y,\rho}|^2)^{1/2}$, this implies

---

[1] The analogous result for minimal surfaces, valid for general stationary varifolds, bounds the "tilt excess" in terms of the mean square deviation of the height of the surface, both quantities measured relative to a given reference plane—see e.g. [SL1] or [AW] for a discussion of such inequalities

(3)
$$|\rho^{2-n} \int_{B_\rho(y)} Dv \cdot D\zeta| \le \sup |\zeta| \delta_0,$$

provided only that

(4)
$$\rho^{-n} \int_{B_\rho(y)} |u - \lambda|^2 \le \delta_0^2,$$

where $\delta_0 \in (0,1)$ will be chosen shortly. Notice also that by using (1) again we see that if $C = C(n,N)$ is chosen suitably here then we also have

(5)
$$\int_{B_{\rho/2}(y)} |Dv|^2 \le 1.$$

On the other hand an easy compactness argument ("harmonic approximation lemma" [SL1]) shows that for any $\epsilon > 0$ there is $\delta_0 = \delta_0(n,\epsilon) > 0$ such that if $v \in W^{1,2}(B_{\rho/2}(y); \mathbf{R}^P)$ is an arbitrary function satisfying (3) and (5) then there is a harmonic function $h = (h^1, \ldots, h^P)$ on $B_{\rho/2}(y)$ such that

(6)
$$\rho^{-n} \int_{B_{\rho/2}(y)} |v - h|^2 \le \epsilon \text{ and } \rho^{2-n} \int_{B_{\rho/2}(y)} |Dh|^2 \le 1.$$

For the simple proof of this we refer to [SL1] or [SL5]. Next note that by elementary calculus along line segments in $B_{\rho/2}(y)$, and by the the standard derivative estimate $\sup_{B_{\rho/4}(y)} |Df|^2 \le C\rho^{-n-2} \int_{B_{\rho/2}(y)} |f|^2$ for harmonic functions f on $B_{\rho/2}(y)$ (which we apply with $f = h - h(y)$), we have
(7)
$$\sup_{B_{\theta\rho}(y)} |h - h(y)|^2 \le \theta^2 \sup_{B_{\theta\rho}(y)} |Dh|^2 \le \theta^2 \sup_{B_{\rho/4}} |Dh|^2 \le C\theta^2 \rho^{-n-2} \int_{B_{\rho/2}(y)} |Dh|^2$$

for any $\theta \in (0, \frac{1}{4})$. Combining (6), (7) and the triangle inequality we conclude
(8)
$$\begin{aligned}
(\theta\rho)^{-n} \int_{B_{\theta\rho}(y)} |v - h(y)|^2 &\le C \sup_{B_{\theta\rho}} |h - h(y)|^2 + C(\theta\rho)^{-n} \int_{B_{\theta\rho}} |v - h|^2 \\
&\le C\theta^2 + C\theta^{-n}\rho^{-n} \int_{B_{\rho/2}} |v - h|^2 \\
&\le C(\theta^2 + \theta^{-n}\epsilon)
\end{aligned}$$

where $C = C(n,N)$, provided (4) holds with suitably small $\delta_0 = \delta_0(n,\epsilon)$. Recalling that $v = C^{-1}(\rho^{-n} \int_{B_\rho(y)} |u - \lambda_{y,\rho}|^2)^{1/2}$, we thus conclude that, for a suitable constant vector $\lambda = C(\rho^{-n} \int_{B_\rho(y)} |u - \lambda_{y,\rho}|^2)^{1/2} h(y)$,

$$(\theta\rho)^{-n} \int_{B_{\theta\rho}(y)} |u - \lambda|^2 \le C(\theta^2 + \theta^{-n}\epsilon)\rho^{-n} \int_{B_\rho(y)} |u - \lambda_{y,\rho}|^2,$$

provided again (4) holds with suitable $\delta_0 - \delta_0(n, \epsilon) > 0$. Since we have $\min_{\lambda \in \mathbf{R}} \int_{B_\sigma(y)} |u - \lambda|^2 = \int_{B_\sigma(y)} |u - \lambda_{y,\sigma}|^2$ for each σ (and in particular for $\sigma = \theta\rho$), we then have

$$(\theta\rho)^{-n} \int_{B_{\theta\rho}(y)} |u - \lambda_{y,\theta\rho}|^2 \leq C(\theta^2 + \theta^{-n}\epsilon)\rho^{-n} \int_{B_\rho(y)} |u - \lambda_{y,\rho}|^2,$$

provided (4) holds with suitable $\delta_0 = \delta_0(n, \epsilon) > 0$. In view of the arbitrariness of θ and ϵ, this evidently implies that

$$(8) \qquad (\theta\rho)^{-n} \int_{B_{\theta\rho}(y)} |u - \lambda_{y,\theta\rho}|^2 \leq \theta^{2\alpha}\rho^{-n} \int_{B_\rho(y)} |u - \lambda_{y,\rho}|^2,$$

for any $\alpha \in (0, 1)$, provided (4) holds with suitable $\delta_0 = \delta_0(n, N, \alpha) > 0$ and $\theta = \theta(\alpha, N) \in (0, 1/2)$. Thus we have "power decay" of the mean square deviation of u away from its average value, provided to begin with the mean square deviation is suitably small as in (4) (with $\delta_0 = \delta_0(n, N, \alpha)$.

In particular (8) implies that (4) again holds with $\theta\rho$ in place of ρ, so we can iterate (8) as many times as we wish, giving

$$(9) \quad (\theta^j\rho)^{-n} \int_{B_{\theta^j\rho}(y)} |u - \lambda_{y,\theta^j\rho}|^2 \leq \theta^{2\alpha}(\theta^{j-1}\rho)^{-n} \int_{B_{\theta^{j-1}\rho}(y)} |u - \lambda_{y,\theta^{j-1}\rho}|^2,$$

for $j = 1, 2, \ldots$, assuming only that (4) holds (at the original radius ρ) with $\delta_0 = \delta_0(n, N, \alpha)$.

Now on the other hand if in place of (4) we assume that

$$(4)' \qquad\qquad \rho^{-n} \int_{B_\rho(y)} |u - \lambda_{y,\rho}|^2 < 2^{-n}\delta_0,$$

then since $\int_{B_{\rho/2}(z)} |u - \lambda_{z,\rho/2}|^2 \leq \int_{B_{\rho/2}(z)} |u - \lambda_{y,\rho}|^2 \leq \int_{B_\rho(y)} |u - \lambda_{y,\rho}|^2$ for any $z \in B_{\rho/2}(y)$, it is easily checked that then $(4)'$ implies that (4) holds with $\rho/2$ in place of ρ and with any $z \in B_{\rho/2}(y)$ in place of y. Hence the above argument actually shows that $(4)'$ implies that (9) holds uniformly with $\rho/2$ in place of ρ and and with any $z \in B_{\rho/2}(y)$ in place of y, and (by a general lemma of Campanato—see e.g. [SL5] for the detailed discussion) this implies that u is Hölder continuous on $B_{\rho/2}(y)$ with Hölder exponent α, and in fact

$$|u(x_1) - u(x_2)| \leq CI^{1/2}(|x_1 - x_2|/\rho)^\alpha,$$

where $I = \rho^{-n} \int_{B_\rho(y)} |u - \lambda_{y,\rho}|^2$.

Thus u is Hölder continuous on $B_{\rho/2}(y)$ subject to $(4)'$. It is then relatively simple to prove the rest of the regularity theory, using standard linear elliptic PDE estimates. (The trickiest step is to prove that u is automatically $C^{1,\alpha}$ once the above Hölder continuity is established; for a discussion of this see e.g. [SL5].

This is as much as we wish to say about the proofs of the ϵ-regularity and compactness theorems here.

The classical Poincaré inequality for $W^{1,2}$ functions on the ball $B_\rho(y)$ guarantees that

$$\rho^{-n} \int_{B_\rho(y)} |u - \lambda_{y,\rho}|^2 \leq C\rho^{2-n} \int_{B_\rho(y)} |Du|^2,$$

where C is a constant depending only on the dimension n, hence the following variant of the ϵ-regularity theorem follows directly from Theorem 1:

Theorem 1'. (*ϵ-regularity theorem—version 2.*) *If* $u \in W^{1,2}(B_\rho(y); N)$ *is energy minimizing and* $\rho^{2-n} \int_{B_\rho(y)} |Du|^2 \leq \epsilon^2$, *with* $\epsilon = \epsilon(n, N) > 0$ *sufficiently small, then* $u \in C^\infty(B_{\rho/2}(y); \mathbf{R}^P)$, *and*

$$\sup_{B_{\rho/2}(y)} \rho^j |D^j u| \leq C_j (\rho^{2-n} \int_{B_\rho(y)} |Du|^2)^{1/2}$$

for each $j = 0, 1, \ldots$.

Now we want to discuss some initial consequences of the ϵ-regularity and compactness theorems.

In this discussion, and subsequently, we let

$$\begin{aligned} \text{reg}\, u &= \{y \in \Omega : u \in C^\infty(B_\sigma(y)) \text{ for some } \sigma > 0\} \\ \text{sing}\, u &= \Omega \backslash \text{reg}\, u. \end{aligned}$$

Notice that reg u is by definition an open subset of Ω and hence sing u is a relatively closed subset of Ω. Our first aim is to show that the ϵ-regularity theorem directly implies that sing u is very small—in fact codimension 2. (Later we show that it is actually codimension 3.)

First we show that the ϵ-regularity theorem gives a nice way of characterizing the regular set:

Corollary 1. $\Theta_u(y) = 0 \iff y \in \text{reg}\, u$.

Proof. "$\Leftarrow$" follows trivially from the fact that u is smooth near y implies $|Du|$ is bounded near y, while "$\Rightarrow$" follows directly from Theorem 1'.

The second corollary shows that the singular set of the energy-minimizing map u is actually quite small:

Corollary 2. $\mathcal{H}^{n-2}(\text{sing}\, u) = 0$. (*In particular* sing $u = \emptyset$ *in case* $n = 2$.)

Remark. Here $\mathcal{H}^j$ is the j-dimensional Hausdorff measure; recall that $\mathcal{H}^j(A) = 0$, for a given subset $A \subset \mathbf{R}^n$, means that for each $\eta > 0$ there is a covering of A by a countable collection $B_{\rho_i}(y_i)$ of balls such that $\sum_i \rho_i^j < \eta$.

Proof. Let K be a compact subset of Ω, $\delta_0 < \text{dist}\,(K, \partial\Omega)$ and $\delta \in (0, \delta_0)$. For $y \in \text{sing}\, u \cap K$ we know by Corollary 2 that

$$(1) \qquad \int_{B_\rho(y)} |Du|^2 \geq \epsilon \rho^{n-2}$$

for all $\rho < \delta_0$. For fixed $\delta < \delta_0$, pick a maximal pairwise disjoint collection $B_{\delta/2}(y_j)_{j=1,\ldots,J}$ for $K \cap \text{sing}\, u$. Then the collection $\{B_\delta(y_j)\}$ covers $K \cap \text{sing}\, u$:

$$(2) \qquad K \cap \text{sing}\, u \subset \cup_j B_\delta(y_j).$$

Using (1) with $\delta/2, y_j$ in place of ρ, y and summing over j we then have (after multiplying through by 2^{n-2})

$$(3) \qquad J\delta^{n-2} \leq \epsilon^{-1} \int_{\cup B_{\rho_j}(y_j)} |Du|^2 \leq \epsilon^{-1} \int_{Q_\delta} |Du|^2,$$

where $Q_\delta = \{x\,:\,\text{dist}\,(x, K \cap \text{sing}\, u) < \delta\}$. In particular

$$J\delta^n \leq \delta^2 \epsilon^{-1} \int_{Q_{\delta_0}} |Du|^2,$$

which, since $B_\delta(y_j)$, $j = 1, \ldots, J$ cover all of $\text{sing}\, u \cap K$, and since we can let $\delta \downarrow 0$, shows that $\text{sing}\, u \cap K$ has Lebesgue measure zero. But then $\int_{Q_\delta} |Du|^2 \to 0$ as $\delta \downarrow 0$ by the dominated convergence theorem, and hence (3) implies that $\mathcal{H}^{n-2}(\text{sing}\, u \cap K) = 0$. Since K was an arbitrary compact subset of $\text{sing}\, u$, this shows that $\mathcal{H}^{n-2}(\text{sing}\, u) = 0$ as required.

Lemma 1. *(Upper semi-continuity of density). If $y_j \to y$, $\rho_0 > 0$, and if $u_j \in W^{1,2}(B_{\rho_0}(y); N)$ is a sequence of energy minimizing maps with $\sup_{j \geq 1} \mathcal{E}_{B_{\rho_0}(y)}(u_j) < \infty$ converging in $L^2(B_{\rho_0}(y); N)$ to u, then $\Theta_u(y) \geq \limsup_{j \to \infty} \Theta_{u_j}(y_j)$.*

Remark. By virtue of the compactness theorem in §2 above, u is automatically energy minimizing and Du_j converges to Du in L^2 on the ball $B_{\rho_0}(y)$.

Proof. If $\rho, \epsilon > 0$ and $\overline{B}_{\rho+\epsilon}(y) \subset \Omega$ then by the above remark, the monotonicity §1(4), and the inclusion $B_\rho(y_j) \subset B_{\rho+\epsilon}(y)$ for sufficiently large j, we have

$$\Theta_{u_j}(y_j) \leq \rho^{2-n} \mathcal{E}_{B_\rho(y_j)}(u_j) \leq \rho^{2-n} \mathcal{E}_{B_{\rho+\epsilon}(y)}(u_j) \to \rho^{2-n} \mathcal{E}_{B_{\rho+\epsilon}(y)}(u).$$

Then the required upper semi-continuity holds by letting $\epsilon \downarrow 0$ and $\rho \downarrow 0$.

LECTURE 2
Tangent Maps and Affine Approximation of Subsets of $\mathbf{R^n}$

1. Tangent Maps

Here u continues to denote an energy minimizing map of Ω into N, with Ω an open subset of $\mathbf{R^n}$.

Take any ball $B_{\rho_0}(y)$ with $\overline{B}_{\rho_0}(y) \subset \Omega$ and for any $\rho > 0$ consider the scaled function $u_{y,\rho}$ defined by

$$u_{y,\rho}(x) = u(y + \rho x).$$

Notice that $u_{y,\rho}$ is well-defined on the ball $B_{\rho_0/\rho}(0)$; furthermore, if $\sigma > 0$ is arbitrary and $\sigma\rho < \rho_0$, we have (using $Du_{y,\rho}(x) = \rho(Du)(y + \rho x)$, and making a change of variable $\tilde{x} = y + \rho x$ in the energy integral for $u_{y,\rho}$)

$$(1) \quad \sigma^{2-n} \int_{B_\sigma(0)} |Du_{y,\rho}|^2 = (\sigma\rho)^{2-n} \int_{B_{\sigma\rho}(y)} |Du|^2 \le \rho_0^{2-n} \int_{B_{\rho_0}(y)} |Du|^2,$$

where in the last inequality we used the monotonicity §1(4) of Lecture 1. Thus if $\rho_j \downarrow 0$ then $\limsup_{j\to\infty} \int_{B_\sigma(0)} |Du_{y,\rho_j}|^2 < \infty$ for each $\sigma > 0$, and hence by the compactness theorem there is a subsequence $\rho_{j'}$ such that $u_{y,\rho_{j'}} \to \varphi$ locally in $\mathbf{R^n}$ both with respect to the L^2-norm and in energy, where $\varphi : \mathbf{R^n} \to N$ is a minimizing harmonic map (in the sense of §1 of Lecture 1 with $\Omega = \mathbf{R^n}$). Any φ which is obtained in this way is called a <u>tangent map of</u> u <u>at</u> y. In general it is <u>not</u> true that such tangent maps need be unique (see [WB])—that is, if we choose different sequences ρ_j (or different subsequences $\rho_{j'}$) then we may get a different limit map.

Note that by the monotonicity identity §1(6) of Lecture 1, with $\rho_{j'}$ in place of ρ (keeping in mind that $u_{y,\rho_{j'}}$ converges in energy to φ) we have, after taking limits on each side of (1) as $j \to \infty$,

$$\sigma^{2-n} \int_{B_\sigma(0)} |D\varphi|^2 = \Theta_u(y),$$

where we used the definition $\Theta_u(y) = \lim_{\rho \downarrow 0} \rho^{2-n} \int_{B_\rho(y)} |Du|^2$. Thus in particular $\sigma^{2-n} \int_{B_\sigma(0)} |D\varphi|^2$ is a constant function of σ, and since $\Theta_\varphi(0) = \lim_{\sigma \downarrow 0} \sigma^{2-n} \int_{B_\sigma(0)} |D\varphi|^2$, we thus have

$$\Theta_u(y) = \Theta_\varphi(0) \equiv \sigma^{2-n} \int_{B_\sigma(0)} |D\varphi|^2 \quad \forall \sigma > 0.$$

Thus any tangent map of u at y has scaled energy constant and equal to the density of u at y; this is also a nice interpretation of the density of u at y.

Furthermore if we apply the monotonicity formula §1(3) of Lecture 1 to φ then we get the identity

$$0 = \sigma^{2-n} \int_{B_\sigma(0)} |D\varphi|^2 - \tau^{2-n} \int_{B_\tau(0)} |D\varphi|^2 = \int_{B_\sigma(0)\backslash B_\tau(0)} r^{2-n} \left|\frac{\partial\varphi}{\partial r}\right|^2,$$

so that $\partial\varphi/\partial r = 0$ a.e., and since φ has L^2 gradient it is correct to conclude from this, by integration along rays, that

(2) $$\varphi(\lambda x) \equiv \varphi(x) \quad \forall \lambda > 0, \, x \in \mathbf{R}^n.$$

This homogeneity is a key property of tangent maps, and we shall capitalize on it below.

2. Properties of Homogeneous Degree Zero Minimizers

Suppose $\varphi \in W^{1,2}(\mathbf{R}^n; N)$ is a homogeneous degree zero minimizer (e.g. a tangent map of u at some point y); thus $\varphi(\lambda x) \equiv \varphi(x)$ for all $\lambda > 0$, $x \in \mathbf{R}^n$.

We first observe that the density $\Theta_\varphi(y)$ is maximum at $y = 0$; in fact by the monotonicity formula §1(6) of Lecture 1, for each $\rho > 0$ and each $y \in \mathbf{R}^n$,

$$2 \int_{B_\rho(y)} r_y^{2-n} \left|\frac{\partial\varphi}{\partial r_y}\right|^2 + \Theta_\varphi(y) = \rho^{2-n} \int_{B_\rho(y)} |D\varphi|^2,$$

where $r_y(x) \equiv |x-y|$ and $\partial/\partial r_y = |x-y|^{-1}(x-y)\cdot D$. Now $B_\rho(y) \subset B_{\rho+|y|}(0)$, so that

$$\rho^{2-n} \int_{B_\rho(y)} |D\varphi|^2 \leq \rho^{2-n} \int_{B_{\rho+|y|}(0)} |D\varphi|^2$$
$$= \left(1 + \frac{|y|}{\rho}\right)^{n-2} (\rho+|y|)^{2-n} \int_{B_{\rho+|y|}(0)} |D\varphi|^2 \equiv \left(1 + \frac{|y|}{\rho}\right)^{n-2} \Theta_\varphi(0),$$

because φ is homogeneous of degree zero (hence $\tau^{2-n} \int_{B_\tau(0)} |D\varphi|^2 \equiv \Theta_\varphi(0)$). Thus letting $\rho \uparrow \infty$, we get

$$2 \int_{\mathbf{R}^n} r_y^{2-n} \left|\frac{\partial\varphi}{\partial r_y}\right|^2 + \Theta_\varphi(y) \leq \Theta_\varphi(0),$$

which establishes the required inequality

(1) $$\Theta_\varphi(y) \leq \Theta_\varphi(0).$$

Notice also that this argument shows that equality can hold in (1) if and only if $\partial\varphi/\partial r_y = 0$ a.e., that is if and only if $\varphi(y + \lambda x) \equiv \varphi(y + x)$ for each $\lambda > 0$. Since we also have (by assumption) $\varphi(x) \equiv \varphi(\lambda x)$ we can then compute for any $\lambda > 0$ and $x \in \mathbf{R}^n$ that

$$\varphi(x) = \varphi(\lambda x) = \varphi(y + (\lambda x - y)) = \varphi(y + \lambda^{-2}(\lambda x - y))$$
$$= \varphi(\lambda(y + \lambda^{-2}(\lambda x - y))) = \varphi(x + ty),$$

where $t = \lambda - \lambda^{-1}$ is an arbitrary real number. So let $S(\varphi)$ be defined by

$$(2) \qquad S(\varphi) = \{y \in \mathbf{R}^n : \Theta_\varphi(y) = \Theta_\varphi(0)\}.$$

Then we have shown that $\varphi(x) \equiv \varphi(x + ty)$ for all $x \in \mathbf{R}^n$, $t \in \mathbf{R}$, and $y \in S(\varphi)$. Combining this with the fact that if $z \in \mathbf{R}^n$ and $\varphi(x + z) \equiv \varphi(x)$ for all $x \in \mathbf{R}^n$, then $\Theta_\varphi(z) = \Theta_\varphi(0)$ (and hence then $z \in S(\varphi)$ by definition of $S(\varphi)$), we conclude

$S(\varphi)$ is a linear subspace of $\mathbf{R}^n$ and $\varphi(x + y) \equiv \varphi(x)$, $x \in \mathbf{R}^n$, $y \in S(\varphi)$.

(Thus φ is invariant under composition with translation by elements of $S(\varphi)$.) Notice of course that

$$(3) \qquad \dim S(\varphi) = n \iff \varphi = \text{const.}$$

Also a homogeneous degree zero map which is not constant clearly cannot be continuous at 0, so we always have $0 \in \text{sing}\,\varphi$ if φ is non-constant, and hence, since $\varphi(x + z) \equiv \varphi(x)$ for any $z \in S(\varphi)$, we have

$$(4) \qquad S(\varphi) \subset \text{sing}\,\varphi$$

for any non-constant homogeneous degree zero minimizer φ.

We note further that then

$$\dim S(\varphi) \leq n - 3.$$

Indeed if $\dim S(\varphi) \geq n - 2$, then by (4) we would have that $\text{sing}\,\varphi$ contains a subspace of dimension $n - 2$, and hence $\mathcal{H}^{n-2}(\text{sing}\,\varphi) = \infty$, contradiction the fact that energy minimizing maps have singular set with zero $(n-2)$-measure, as shown in Lecture 1 above.

3. Approximation of Subsets of $\mathbf{R}^n$ by Affine Subspaces

We want to devote this section to discussion of some properties of subsets $A \subset \mathbf{R}^n$ which are approximable in various senses by j-dimensional affine spaces.

Definition: Let $A \subset \mathbf{R}^n$ be an arbitrary set and $\epsilon > 0$; then

  (i) A has the <u>weak</u> j-dimensional ϵ-approximation property if for each $y \in A$ there is $\rho_y > 0$ such that $B_\rho(y) \cap A \subset$ the $\epsilon\rho$-neighbourhood of some j-dimensional subspace $L_{y,\rho}$ $\forall \rho \in (0, \rho_0]$.

  (ii) A has the <u>strong</u> j-dimensional ϵ-approximation property if this holds with $L_{y,\rho}$ independent of ρ.

(iii) A has the <u>very strong</u> j-dimensional ε-approximation property if this holds with $L_{y,\rho} = y + L$ $\forall y \in A$, with L a j-dimensional subspace which is independent of both ρ and y

In each case the property is called ρ_0-uniform if the property holds with $\rho_y \equiv \rho_0$.

Concerning these properties, we have the following lemma in which we continue to use the terminology that "G is the graph of a Lipschitz function over some j-dimensional subspace" to mean that there is a j-dimensional subspace $L \subset \mathbf{R}^n$ and a map $u : L \to L^\perp$ such that $\sup_{x,y \in L,\, x \neq y} |x - y|^{-1}|u(x) - u(y)| < \infty$.

Lemma 1. **(i)** *There is a function $\beta : (0,1] \to (0,1]$ with $\lim_{\epsilon \downarrow 0} \beta(\epsilon) = 0$ such that if $A \subset \mathbf{R}^n$ has the j-dimensional weak ε-approximation property for some given $\epsilon \in (0,1]$, then $\mathcal{H}^{j+\beta(\epsilon)}A = 0$. (In particular, if A has the j-dimensional weak ε-approximation property $\forall \epsilon > 0$, then $\dim A \leq j$.)*

(ii) *If A has the strong j-dimensional ε-approximation property for some $\epsilon \in (0,1]$, then $A \subset$ the countable union of Lipschitz graphs.*

(iii) *If $A \subset B_{\rho_0}(x_0)$ has the ρ_0-uniform very strong j-dimensional ε-approximation property for some $\epsilon \in (0,1]$, then $A \subset \operatorname{graph} f$, the graph of a Lipschitz function f over some j-dimensional subspace of $\mathbf{R}^n$ with Lipschitz constant ε.*

Remark. (1) There are simple examples to show that the result in part (i) above cannot be improved. For example the family of "Koch" curves from fractal geometry includes sets which satisfy the uniform 1-dimensional ε-approximation property but which have dimension > 1. For a more complete discussion of this, see [SL7].

(2) Notice that the ρ_0-uniform very strong ε-approximation assumed in (iii) requires that there is a j-dimensional subspace L such that

$$A \cap B_\rho(y) \subset (\epsilon\rho)\text{-neighbourhood of } y + L$$

for every $y \in A$ and $\rho \leq \rho_0$; this is a "uniform cone condition" on the set A and the conclusion of (iii) is standard. Notice that if we strengthen the hypothesis to require that there is a map $\sigma : (0,1] \to (0,\epsilon]$ which is increasing and $\lim_{t \downarrow 0} \sigma(t) = 0$, then we can make the stronger conclusion that $A \subset \Sigma$, where Σ is the graph of a C^1 function (with values in $L^\perp$) over L.

(3) The property that A is contained in the countable union of Lipschitz graphs is known as "countable j-rectifiability", so in fact the conclusion of (ii) says that A is countably j-rectifiable.

The proof of (i) of the above lemma is an easy exercise based on the definition of Hausdorff measure, using the given weak approximation property to construct successively finer coverings of A by balls. See e.g. [SL7]. The result of (iii) follows, in view of Remark (2) above, from the characterization of Lipschitz graphs as those subsets which satisfy a uniform cone condition of the type mentioned in Remark (2) above. For the detailed argument, we refer to e.g. [SL7] or [SL4]. Finally (ii) easily follows from (iii) after making the subdivision of A into a suitable countable subcollection. Again the details can be found in e.g. [SL7] or [SL4].

4. Further Properties of sing u

We define the subset sing_*u of sing u to be the set of points $y \in \text{sing } u$ such that at least one tangent map φ of u at y has $\dim S(\varphi) = n - 3$.

Thus sing_*u is the subset of sing u where there is a tangent map φ such that $S(\varphi)$ has maximum dimension.

We call sing_*u the "top dimensional part" of sing u.

The following lemma shows that sing_*u accounts for "most" of the singular set (unless the singular set already has dimension $\leq n - 4$).

Theorem 1. *Let $u \in W^{1,2}(\Omega; N)$ be energy minimizing. Then*

$$\dim \text{sing } u \leq n - 3$$

and

$$\dim(\text{sing } u \backslash \text{sing}_*u) \leq n - 4.$$

In [SU], Schoen and Uhlenbeck used Federer's "dimension reducing" argument to establish the first part of this lemma. Here we use a refinement of this argument, due to Almgren (who used the argument in the context of area minimizing currents in [A1]), which yields also the second conclusion.

For the proof of the first part we need the following lemma:

Lemma 2. *If $u \in W^{1,2}(\Omega; N)$ is energy minimizing, if $y \in \text{sing} u$, and if $\epsilon > 0$ then there are $\rho(y, u, \epsilon) > 0$, $\eta = \eta(y, u, \epsilon) > 0$ such that for each $\rho \in (0, \rho(y, u, \epsilon)]$*

$$\{x \in B_\rho(y) : \Theta_u(x) \geq \Theta_u(y) - \eta\} \subset \text{ the } (\epsilon\rho)\text{-neighbourhood of } L_{y,\rho}$$

for some $(n-3)$-dimensional affine space $L_{y,\rho}$ of $\mathbf{R}^n$ (depending on y, ρ).

Proof. If this is false for some $\epsilon > 0$, there is a sequence $\rho_j \downarrow 0$ such that

$$\{x \in B_{\rho_j}(y) : \Theta_u(x) \geq \Theta_u(y) - 1/j\} \not\subset \text{ the } (\epsilon\rho_j)\text{-neighbourhood of } L$$

for each $(n-3)$-dimensional affine space L. Notice that by scaling, this says

$$(1) \quad \{x \in B_1(0) : \Theta_{u_{y,\rho_j}}(x) \geq \Theta_u(y) - 1/j\} \not\subset \text{ the } (\epsilon\rho)\text{-neighbourhood of } L$$

for each $(n-3)$-dimensional affine space L. Now by the above discussion of tangent maps, we know there is a subsequence $\rho_{j'}$ such that

$$u_{y,\rho_{j'}} \to \varphi,$$

where φ is a tangent map of u at y, so that $\Theta_\varphi(0) = \Theta_u(y)$. Now by our discussion in §2 above, we have an $(n-3)$-dimensional subspace $L \subset \mathbf{R}^n$ such that

$$\{x \in B_1(0) : \Theta_\varphi(x) \geq \Theta_u(y)\} \subset L.$$

Using the upper semi-continuity of Θ discussed in Lemma 1 of Lecture 1, we easily check that this contradicts (1).

The proof of the second part of the theorem will require the the following:

Lemma 3. *If $u \in W^{1,2}(\Omega; N)$ is energy minimizing, if $y \in \text{sing } u \backslash \text{sing}_* u$, and if $\epsilon > 0$ then there are $\rho_0 = \rho_0(y, u, \epsilon) > 0$, $\eta = \eta(y, u, \epsilon) > 0$ such that for each $\rho \in (0, \rho_0]$*

$$\{x \in B_\rho(y) : \Theta_u(x) \geq \Theta_u(y) - \eta\} \subset \text{ the } (\epsilon\rho)\text{-neighbourhood of } L_{y,\rho}$$

for some $(n-4)$-dimensional affine space $L_{y,\rho}$ of $\mathbf{R}^n$ (depending on y, ρ).

The proof of this lemma is identical to the proof of Lemma 2, except that we substitute the dimension $n-4$ wherever dimension $n-3$ appears in the proof of Lemma 2.

Now we can prove Theorem 1. The first part is proved by simply decomposing

(1) $$\text{sing } u = \cup_k^\infty S_{ik},$$

for each $i = 1, 2, \ldots$, where S_{ik} denotes the subset of sing u consisting of those points y such that the conclusion of Lemma 2 holds with $\epsilon = 1/i$, $\eta \geq 1/k$, $\rho(y, u, 1/i) \geq 1/k$ and such that $\Theta_u(y) \in ((i-1)/k, i/k]$. (Notice that such a decomposition is guaranteed by Lemma 2.) On the other hand by Lemma 2 it is now easy to check that each S_{ik} has the weak $(n-3)$-dimensional $(1/i)$-approximation property of §3 above, and so by Lemma 1(i) we have $\mathcal{H}^{n-3+\epsilon_i}(S_{ik}) = 0$ for each i, k, where $\epsilon_i \downarrow 0$ as $i \to \infty$. Thus by (1) $\mathcal{H}^{n-3+\epsilon_i}(\text{sing } u) = 0$ for each i, and hence dim sing $u \leq n-3$ as required.

The proof of the second part of the theorem is identical, except that we use $(n-4)$-dimensional subspaces instead of $(n-3)$ and Lemma 3 instead of Lemma 2.

We conclude this section with an important observation about weak ϵ-approximation property of certain parts of the singular set.

Lemma 4. *If $u \in W^{1,2}(\Omega; N)$ is energy minimizing, if $\epsilon > 0$, and if $y \in \text{sing } u$, then there is $\rho_0 = \rho_0(y, u) > 0$ such that the set*

$$\{x \in \text{sing } u \cap B_{\rho_0}(y) : \Theta_u(x) \geq \Theta_u(y)\}$$

has the ρ_0-uniform weak $(n-3)$-dimensional ϵ-approximation property.

The proof is almost identical to the proof of Lemma 2 above, except that we need to use pseudo-tangent maps in place of the tangent maps φ. (A pseudo-tangent map of u at y is a map $\varphi = \lim u_{y_j, \rho_j}$, where $y_j \to y$, $\rho_j \downarrow 0$, and $\Theta_u(y_j) \geq \Theta_u(y) \, \forall j$; a simple modification of the discussion of tangent maps in §1 above shows that such φ are still homogeneous of degree zero). We refer to [SL7] or [SL8] for the complete argument.

5. Homogeneous Degree Zero φ with $\dim S(\varphi) = n - 3$

Let $\varphi : \mathbf{R}^n \to N$ be any homogeneous degree zero minimizer with $\dim S(\varphi) = n - 3$. Then, modulo a rotation of the x-variables which takes $S(\varphi)$ to $\{0\} \times \mathbf{R}^{n-3}$, we have

$$(1) \qquad\qquad \varphi(x, y) \equiv \varphi_0(x),$$

where we use the notation that a general point in $\mathbf{R}^n$ is denoted (x, y), $x \in \mathbf{R}^3$, $y \in \mathbf{R}^{n-3}$, and where φ_0 is a homogeneous degree zero map from $\mathbf{R}^3$ into N. We in fact claim that

$$(2) \qquad\qquad \operatorname{sing} \varphi_0 = \{0\} \quad \text{and hence } \varphi_0 | S^2 \in C^\infty,$$

so that $\varphi_0 | S^2$ is a smooth harmonic map of S^2 into N. To see this, first note that $\operatorname{sing} \varphi_0 \supset \{0\}$, otherwise φ_0, and hence φ, would be constant, thus contradicting the hypothesis $\dim S(\varphi) = n - 3$. On the other hand if $\xi \neq 0$ with $\xi \in \operatorname{sing} \varphi_0$, then by homogeneity of φ_0 we would have $\{\lambda \xi : \lambda > 0\} \subset \operatorname{sing} \varphi_0$, and hence

$$\{(\lambda \xi, y) : \lambda > 0, \, y \in \mathbf{R}^{n-3}\} \subset \operatorname{sing} \varphi.$$

But the left side here is a half-space of dimension $(n-2)$, and hence this would give $\mathcal{H}^{n-2}(\operatorname{sing} \varphi) = \infty$, thus contradicting the fact that $\mathcal{H}^{n-2}(\operatorname{sing} \varphi) = 0$ by §2 of Lecture 1. Thus (2) is established.

Thus the smooth harmonic maps $S^2 \to N$ are of particular significance in the study of the top dimensional part of the singular set of an arbitrary energy minimizing map. We shall have more to say about them in the next lecture.

LECTURE 3
Asymptotics on Approach to Singular Points

1. Significance of Unique Asymptotic Limits

With $u \in W^{1,2}(\Omega; N)$ still denoting an energy minimizing map, let $y \in$ sing$_*\, u$. By definition of sing$_*\, u$ in the previous lecture there is a tangent map φ of u at y with dim $S(\varphi) = n - 3$. As in §4 of Lecture 2, we assume without loss of generality (after making an orthogonal transformation of the x variables which takes $S(\varphi)$ to $\{0\} \times \mathbf{R}^{n-3}$), that

$$(1) \qquad \varphi(x, y) \equiv \varphi_0(x), \qquad x \in \mathbf{R}^3,\, y \in \mathbf{R}^{n-3}.$$

By definition of tangent map, there is a sequence $\rho_j \downarrow 0$ such that

$$(2) \qquad \lim_{j \to \infty} \rho_j^{-n} \int_{B_{\rho_j}(y)} |u - \varphi|^2 = 0.$$

Since sing $\varphi = \mathbf{R}^{n-3} \times \{0\}$, so that φ is smooth on $\mathbf{R}^n \backslash (\mathbf{R}^{n-3} \times \{0\})$, it is then clear from the ϵ-regularity theorem that the convergence of u_{y,ρ_j} to φ is actually in the C^k-norm, for any k, locally in the region $\mathbf{R}^n \backslash (\mathbf{R}^{n-3} \times \{0\})$, and in particular for each $\epsilon > 0$ we have

$$(3) \qquad B_\rho \cap \text{sing}\, u \subset \text{ the } (\epsilon\rho)\text{-neighbourhood of } S(\varphi)$$

for $\rho = \rho_j$ with j sufficiently large depending on ϵ. Notice in particular that if the tangent map φ were the <u>unique</u> tangent map of u at y, so that $\lim_{\rho \downarrow 0} u_{y,\rho} = \varphi$ in $L^2_{\text{loc}}(\mathbf{R}^n; \mathbf{R}^P)$, then we would have (3) for all sufficiently small ρ and all $y \in$ sing$_*\, u$. We would therefore have the strong $(n - 3)$-dimensional ϵ-approximation property for sing$_*\, u$, and hence, by Lemma 1 of §2 of Lecture 2, we would have the countable rectifiabilty of sing u. Unfortunately the uniqueness of tangent maps is false in general (see the discussion in [SL7] and the explicit counterexample in [WB]), so we cannot use this method to establish rectifiability results about the singular set.

On the other hand, uniqueness of tangent maps is still very much an open question in the case when the target manifold N is real-analytic, and indeed in this case uniqueness is known at points y at which u has at least one tangent map φ with sing $\varphi = \{0\}$:

Theorem 1 ([SL6]). *Suppose $u \in W^{1,2}(\Omega; N)$ is energy minimizing and N is real-analytic, and suppose that there is a tangent map φ of u at a point $y \in$ sing u with sing $\varphi = \{0\}$. Then φ is the unique tangent map of u at y.*

Since it is relevant to the discussion in the next lecture, we do want briefly to outline a recent simplification of the proof of the above theorem. We first need some facts about the energy functional which follow from a theorem of Lojasiewicz for real analytic functionals.

2. Lojasiewicz Inequalities for the Energy

We here need a "Lojasiewicz type" inequality for the energy functional on spheres; we begin by noting that, according to Lojasiewicz [L], if f is a real-analytic function on some open set U of some Euclidean space $\mathbf{R}^Q$, then for each critical point $y \in U$ of f (i.e., each point y where $\nabla f(y) = 0$) there is $\alpha \in (0,1]$ and $\sigma > 0$ such that

$$(1) \qquad |\nabla f(x)| \geq |f(x) - f(y)|^{1-\alpha/2}$$

for every point $x \in B_\sigma(y)$.

There is an infinite dimensional analogue of this inequality which applies to the energy functional $\mathcal{E}_{S^{\ell-1}}$. Specifically, for any $\varphi_0 \in C^\infty(S^{\ell-1}; N)$ which is stationary for the energy functional (in the sense that the equation $\Delta\varphi_0 + \sum_{j=1}^{n-1} A_{\varphi_0}(D_j\varphi_0, D_j\varphi_0) = 0$ holds for φ_0 (Cf. §1(1)' of Lecture 1), where we use the homogeneous degree zero extension of φ_0 in computing $D_j\varphi_0$), then there is $\alpha = \alpha(\varphi_0, n, N) \in (0,1)$ and $\sigma = \sigma(\varphi_0, n, N) > 0$ such that

$$(2) \qquad |\mathcal{E}_{S^{\ell-1}}(\psi) - \mathcal{E}_{S^{\ell-1}}(\varphi_0)|^{1-\alpha/2} \leq \|(\Delta\psi)^T\|_{L^2},$$

where the notation is as in §1 of Lecture 1.

We shall not give the proof of inequality (2) here (it can be found in [SL6] or [SL7]), but we do want to point out that it follows from the finite dimensional inequality (1) via a construction (known as the Liapunov-Schmidt reduction) which essentially reduces the study of the set of solutions of the elliptic Euler-Lagrange equation of a general functional $\mathcal{F}$ on a compact manifold Σ to the study of the local set of critical points of a function on some open set in a finite dimensional Euclidean space. We sketch this argument for the case when the functions u are scalar-valued, but only notational changes are needed to handle the vector-valued case; for complete details, including the vector case, we refer to [SL6] or [SL7].

To describe this, let $\mathcal{F}(v) = \int_\Sigma F(\omega, v, \nabla v)$ be a functional on a compact Riemannian manifold Σ, where F is smooth and has real analytic dependence on v, ∇v. We let $\mathcal{L}_\mathcal{F} v \equiv \frac{d}{ds}\mathcal{F}(sv)|_{s=0}$ denote the linearization of the Euler-Lagrange operator $\mathcal{M}_\mathcal{F}$ at 0 and let K denote the (finite-dimensional) kernel of this linear operator; $\mathcal{L}_\mathcal{F}$ is elliptic by definition because we assume that $\mathcal{M}_\mathcal{F}$ is elliptic.

Let Π denote the $L^2(\Sigma)$-orthogonal projection of $L^2(\Sigma)$ onto K, and note that the nonlinear operator $\mathcal{N} \equiv \Pi + \mathcal{M}_\mathcal{F}$ has linearization (on the space $C^2(\Sigma)$) equal to $\Pi + \mathcal{L}_\mathcal{F}$, which evidently has trivial kernel, so the Schauder estimates for linear elliptic equations guarantee we can apply the inverse function theorem to give an inverse $\Psi : U \to V$, where U, V denote neighbourhoods of 0 in $C^{0,\alpha}(\Sigma)$ and $C^{2,\alpha}(\Sigma)$ respectively. We can also define a scalar function $f = \mathcal{F} \circ \Psi$ on W. Then an easy computation shows that (provided the neighbourhood U is sufficiently small) we have

$$(3) \qquad |\nabla f(\xi)| \leq C\|\mathcal{M}_\mathcal{F}(\Psi(\xi))\|, \quad \xi \in K \cap U,$$

and also

(4) $\{u \in U \;:\; \mathcal{M}_{\mathcal{F}}(u) = 0\} = \Psi(\{\xi \in W \;:\; \nabla f(\xi) = 0\}),$

provided U, W are appropriately selected neighbourhoods of 0 in $C^{2,\alpha}(\Sigma)$ and K respectively.

Hence in particular the set of $u \in U$ such that $\mathcal{M}_{\mathcal{F}}(u) = 0$ is contained in the finite dimensional manifold $M = \Psi(W)$. It is also straightforward to prove the inequalities

(5) $|f(\xi) - \mathcal{F}(u)| \leq C\|\mathcal{M}_{\mathcal{F}}(u)\|^2_{L^2(\Sigma)}, \quad \xi = \Pi(u),\ \xi \in W,\ u \in U.$

In case the function F used in the definition of the functional $\mathcal{F}$ has real-analytic dependence on u, ∇u, the funcion f is also real-analytic, and hence satisfies a Lojasiewicz inequality as in (1) above. It then follows from (3) and (5) that we have the Lojasiewicz inequality

$$|\mathcal{F}(u) - \mathcal{F}(0)|^{1-\alpha/2} \leq C\|\mathcal{M}_{\mathcal{F}}(u)\|, \quad u \in \widetilde{U}$$

for $\widetilde{U}$ a suitable neighbourhood of 0 in $C^{2,\mu}(\Sigma)$. Even in the general case when F is merely smooth we note the important fact that

(6) $f \equiv$ const. in some neighbourhood of 0

implies the Lojasiewicz inequality with best exponent $\alpha = 1$. To see this we simply note that if $f \equiv$ const., then

$$|\mathcal{F}(u) - \mathcal{F}(0)| \equiv |\mathcal{F}(u) - f(\xi)| \leq C\|\mathcal{M}_{\mathcal{F}}(u)\|^2_{L^2}, \quad \xi = \Pi(u), u \in U, \xi \in W,$$

by virtue of (5) above. (6) is called an "integrablity condition" because by (4) it implies that the set of solutions of the non-linear equation $\mathcal{M}_{\mathcal{F}}(u) = 0$ in a small enough neighbourhood of 0 forms a finite dimensional manifold with the same dimension as the dimension of the kernel (Viz., the manifold M above).

Using the corresponding version of these relations in a suitable vector-bundle setting, the Lojasiewicz inequality (2) for the energy funcional follows. Further we have the same inequality with best exponent $\alpha = 1$, assuming that an "integrability condition" like (6) holds for the energy functional with φ as origin; that is, assuming that, in a C^3-neighbourhood of φ, the set of smooth harmonic maps S^m into N forms a manifold with dimension equal to the dimension of the kernel of the linearization of the operator $\triangle_{S^m}\varphi + \sum_{i=1}^n A_\varphi(D_i\varphi, D_i\varphi) = 0$. That is, if N is merely smooth, then assuming (6) for the energy functional with origin at φ_0, there are C, $\sigma > 0$ such that

(7) $|\mathcal{E}_{S^{n-1}}(u) - \mathcal{E}_{S^{n-1}}(\varphi_0)|^{1/2} \leq C\|\mathcal{M}_{\mathcal{E}_{S^{n-1}}}(u)\|_{L^2}$

whenever $u \in C^\infty(S^{n-1}; N)$ with $\|u - \varphi_0\|_{C^3} < \sigma$.

Remark: If $\dim N = 2$ the integrability condition automatically holds for any smooth harmonic map $\varphi_0 \in C^\infty(S^2; N)$, as shown in [GW]. The integrabilty also holds if N is metrically sufficiently close to S^3 in the C^3 sense: see the discussion in [SL3].

3. Proof of Theorem 1 of §1

First recall, by definition of tangent map, that there is a sequence $\rho_j \downarrow 0$ such that the rescaled mapping $u_{y,\rho_j}(= u(y + \rho_j x))$ converge in the $W^{1,2}$-norm to our tangent map φ. Thus for any given $\eta > 0$, for suitable ρ, for example for $\rho = \rho_j$ with j sufficiently large depending on η, we have in particular that

$$(1) \qquad \int_{B_{3/2} \setminus B_{3/4}} |u_{y,\rho} - \varphi|^2 < \eta^2,$$

where, here and subsequently, B_ρ is an abbreviation for $B_\rho(0)$. Let us now abbreviate $\tilde{u} = u_{y,\rho}$, and keep this ρ fixed for the time being and small enough so that $\overline{B}_{3\rho/2} \subset \Omega$, so that $\tilde{u}$ is at least defined on $\overline{B}_{3/2}$. Now since φ is smooth in $B_{3/2} \setminus B_{3/4}$, it is clear that if $B_\sigma(z) \subset B_{3/2} \setminus B_{3/4}$ then

$$\sigma^{-n} \int_{B_\sigma(z)} |\tilde{u} - \varphi(z)|^2 \;\leq\; 2\sigma^{-n} \int_{B_\sigma(z)} |\tilde{u} - \varphi|^2 + 2\sigma^{-n} \int_{B_\sigma(z)} |\varphi - \varphi(z)|^2$$
$$\leq\; 2\sigma^{-n}\eta^2 + \beta\sigma^2,$$

where β is a fixed constant depending on φ but not depending on σ or ρ. Thus if $\gamma > 0$ is given, then for small enough η, σ (depending only on n, N, φ, γ) we can apply the ϵ-regularity theorem on the ball $B_\sigma(z)$ in order to deduce that $\|u - \varphi\|_{C^3(B_{\sigma/2}(z))} \leq \gamma$. Thus (in view of the arbitrariness of z) we obtain for any given $\gamma > 0$ that there exists $\eta = \eta(\gamma, \varphi) > 0$ such that

$$(2) \qquad \|\tilde{u} - \varphi\|_{L^2(B_{3/2}\setminus B_{3/4})} < \eta \Rightarrow \|\tilde{u} - \varphi\|_{C^3(B_{5/4}\setminus B_{7/8})} < \gamma.$$

(Notice that in (2) we do <u>not</u> have to assume that ρ is proportionately close to one of the ρ_j.)

Next recall that by §2(2) we have constants $C > 0$, $\gamma \in (0,1)$ and $\alpha \in (0,1]$ such that

$$(3)$$
$$|\int_{S^{n-1}} (|D_\omega w|^2 - |D_\omega \varphi|^2)|^{1-\alpha/2} \leq C\|\mathcal{M}_{\mathcal{E}_{S^{n-1}}}(w)\|_{L^2}, \qquad \|w - \varphi\|_{C^3} < \gamma.$$

Since $\|\mathcal{M}_{\mathcal{E}_{S^{n-1}}}(w)\|_{L^2}$ is uniformly bounded for $\|w\|_{C^2} \leq 1$ (so that then trivially $\|\mathcal{M}_{\mathcal{E}_{S^{n-1}}}(w)\|_{L^2}^{\gamma_2} \leq C^{\gamma_2 - \gamma_1}\|\mathcal{M}_{\mathcal{E}_{S^{n-1}}}(w)\|_{L^2}^{\gamma_1}$ for any $\gamma_1 \leq \gamma_2$), we can, and we shall, assume $\alpha < 1$.

So from now on $\alpha \in (0,1), \gamma$ (depending on φ) are chosen fixed such that (3) holds, and η, depending on φ and γ, is chosen so that the implication (2) holds.

Now by the monotonicity identity §1(6) of Lecture 1 we have

$$(4) \qquad 2\int_{B_1} r^{2-n}|\frac{\partial \tilde{u}}{\partial r}|^2 = \int_{B_1} |D\tilde{u}|^2 - \Theta_{\tilde{u}}(0).$$

Also, in proving this identity (in §2.4) we showed that $\tilde{u}$ satisfied the identity

$$(n-2)\int_{B_1}|D\tilde{u}|^? = \int_{\partial B_1}(|D\tilde{u}|^2 - 2|\frac{\partial\tilde{u}}{\partial r}|^2) \leq \int_{\partial B_1}|D'\tilde{u}|^2,$$

where D' means tangential gradient on ∂B_1. Then using this in (4), and keeping in mind that

$$\Theta_{\tilde{u}}(0) = \Theta_u(y) = \Theta_\varphi(0) = \int_{B_1}|D\varphi|^2 = \frac{1}{n-2}\int_{S^{n-1}}|D_\omega\varphi|^2$$

by virtue of the fact that φ is a tangent map of u at y (hence homogeneous of degree zero), we obtain

(5)
$$2(n-2)\int_{B_1}r^{2-n}|\frac{\partial\tilde{u}}{\partial r}|^2 \leq \int_{\partial B_1}(|D'\tilde{u}|^2 - |D'\varphi|^2)$$
$$= \int_{S^{n-1}}(|D_\omega\tilde{u}|^2 - |D_\omega\varphi|^2).$$

Now in view of (2) we can apply the inequality (3) in order to deduce

(6)
$$\int_{B_1}r^{2-n}|\frac{\partial\tilde{u}}{\partial r}|^2 \leq C\|\mathcal{M}(\tilde{u})\|_{L^2}^{1/(1-\alpha/2)},$$

so long as $\|\tilde{u} - \varphi\|_{L^2(B_{3/2}\backslash B_{3/4})} < \eta$.

Now $\tilde{u}$ satisfies the equation $\Delta\tilde{u} + \sum_{j=1}^n A_{\tilde{u}}(D_j\tilde{u}, D_j\tilde{u}) = 0$, which, in terms of spherical coordinates $r = |x|$, $\omega = |x|^{-1}x$, can be written

$$\frac{1}{r^{n-1}}\frac{\partial}{\partial r}(r^{n-1}\frac{\partial\tilde{u}}{\partial r}) + \frac{1}{r^2}\Delta_{S^{n-1}}\tilde{u} + \frac{1}{r^2}A_{\tilde{u}}(D_\omega\tilde{u}, D_\omega\tilde{u}) + A_{\tilde{u}}(\frac{\partial\tilde{u}}{\partial r}, \frac{\partial\tilde{u}}{\partial r}) = 0.$$

Hence (6) implies

(7)
$$(\int_{B_1}r^{2-n}|\frac{\partial\tilde{u}}{\partial r}|^2)^{2-\alpha} \leq C\int_{S^{n-1}}(|\frac{\partial(r^{n-1}\partial\tilde{u}/\partial r)}{\partial r}|^2 + |\frac{\partial\tilde{u}}{\partial r}|^2),$$

provided that $\|\tilde{u} - \varphi\|_{L^2(B_{3/2}\backslash B_{3/4})} < \eta$.

Now notice that the rescaled function $\tilde{u}(\sigma)$ defined by $\tilde{u}(\sigma)(x) = \tilde{u}(\sigma x)$ also satisfies the harmonic map equation

$$\Delta\tilde{u}(\sigma) + \sum_{j=1}^n A_{\tilde{u}(\sigma)}(D_j\tilde{u}(\sigma), D_j\tilde{u}(\sigma)) = 0,$$

and by differentiation with respect to σ, and noting that

$$\frac{\partial\tilde{u}(\sigma x)}{\partial\sigma}|_{s=1} = x \cdot D\tilde{u}(x) = r\frac{\partial\tilde{u}}{\partial r},$$

we obtain the linear equation

$$\mathcal{L}(r\frac{\partial\tilde{u}}{\partial r}) = 0,$$

where $\mathcal{L}$ is the linearization of $\mathcal{M}(u)$ at $\tilde{u}$; thus $\mathcal{L}$ is the linear elliptic operator

$$\mathcal{L}v = \Delta v + 2\sum_{j=1}^{n} A_{\tilde{u}(\sigma)}(D_j v, D_j \tilde{u}(\sigma)) + \sum_{j=1}^{n}\sum_{k=1}^{p} v^k \frac{\partial A_z}{\partial z^k}\Big|_{z=\tilde{u}(\sigma)}(D_j\tilde{u}(\sigma), D_j\tilde{u}(\sigma)).$$

Now since $\|\tilde{u} - \varphi\|_{C^3(B_{5/4}\backslash B_{7/8})} \leq \gamma \leq 1$ by (2), we see that this operator has the form

$$\mathcal{L}(v) = \Delta v + b \cdot Dv + c \cdot v,$$

where $|b| + |c| \leq \beta$ in the domain $\Omega = B_{5/4}\backslash B_{7/8}$. But any solution $v = (v^1, \ldots, v^p)$ of $\mathcal{L}(v) = 0$ for such an operator $\mathcal{L}$ satisfies the estimate

$$\sup_{B_{\tau/2}(z)} |Dv| \leq C\|v\|_{L^2(B_\tau(z))}$$

for any ball $B_\tau(z) \subset \Omega(= B_{5/4}\backslash B_{7/8})$. (See e.g. [GT].) Thus in particular covering S^{n-1} by a family of such balls $B_{\tau/2}$ with $\tau = 1/16$, we conclude

$$\sup_{S^{n-1}} |D(r\frac{\partial \tilde{u}(\sigma)}{\partial r})|^2 \leq C\int_{B_{5/4}\backslash B_{7/8}} |\frac{\partial \tilde{u}(\sigma)}{\partial r}|^2,$$

provided that $\|\tilde{u} - \varphi\|_{L^2(B_{3/2}\backslash B_{3/4})} < \eta$. Using the definition (6) and replacing σ by $\sigma/2$ we have

$$(\int_{B_1} r^{2-n}|\frac{\partial \tilde{u}}{\partial r}|^2)^{2-\alpha} \leq C\int_{B_{3/2}\backslash B_{3/4}} r^{2-n}|\frac{\partial \tilde{u}}{\partial r}|^2,$$

provided $\|\tilde{u} - \varphi\|_{L^2(B_{3/2}\backslash B_{3/4})} < \eta$. By rescaling (i.e. applying the argument with $\tilde{u}(\sigma x)$ in place of $\tilde{u}(x)$), since $\frac{2}{3} \cdot \frac{3}{4} = \frac{1}{2}$ we in fact deduce

$$(8) \qquad (\int_{B_{\sigma/2}} r^{2-n}|\frac{\partial \tilde{u}}{\partial r}|^2)^{2-\alpha} \leq C\int_{B_\sigma\backslash B_{\sigma/2}} r^{2-n}|\frac{\partial \tilde{u}}{\partial r}|^2$$

for <u>any</u> $\sigma \in (0, \frac{3}{2}]$ such that $\sigma^{-n}\|\tilde{u} - \varphi\|^2_{L^2(B_\sigma\backslash B_{\sigma/2})} < \eta^2$, where C is a constant depending only on φ. This is the key inequality; we claim that the theorem now follows quite easily from it.

To see this, let $I(\sigma) = \int_{B_\sigma} r^{2-n}|\frac{\partial \tilde{u}}{\partial r}|^2$, and note that by (5) and (2) that we have all $I(\sigma) \leq C\gamma$ so long as $\sigma^{-n}\|\tilde{u} - \varphi\|^2_{L^2(B_\sigma\backslash B_{\sigma/2})} < \eta^2$, so that, by making a smaller choice of γ from the start if necessary, we can assume that $I(\sigma) \leq 1$ for each σ such that $\sigma^{-n}\|\tilde{u} - \varphi\|^2_{L^2(B_\sigma\backslash B_{\sigma/2})} < \eta^2$. Next we note that

(9)
$$0 < a < b \leq 1, \alpha \in (0,1), \beta > 0, \text{ and } a^{2-\alpha} \leq \beta(b-a) \Rightarrow a^{\alpha-1} - b^{\alpha-1} \geq C,$$

where C is a fixed constant determined by α, β only. This is readily checked by calculus, considering separately the cases when $b/a \geq 2$ and $b/a < 2$.

Notice that in view of (8) we can apply (9) with $a = I(\sigma/2)$ and $b = I(\sigma)$, thus giving

(10) $$I(\sigma/2)^{\alpha-1} - I(\sigma)^{\alpha-1} \geq C,$$

provided $\sigma^{-n}\|\widetilde{u} - \varphi\|^2_{L^2(B_\sigma \backslash B_{\sigma/2})} < \eta^2$. If we use the notation that $\widetilde{u}(\sigma)$ denotes the function $\in C^\infty(S^{n-1}; N)$ given by $\widetilde{u}(\sigma)(\omega) = \widetilde{u}(\sigma\omega)$, then $\int_{B_{\rho_2}\backslash B_{\rho_1}} |\widetilde{u}-\varphi|^2 = \int_{\rho_1}^{\rho_2} \sigma^{n-1}\|\widetilde{u}(\sigma)-\varphi\|^2_{L^2(S^{n-1})} d\sigma$, hence certainly (10) holds if we require

(11) $$\|\widetilde{u}(\tau) - \varphi\|_{L^2(S^{n-1})} < \eta, \qquad \forall\, \tau \in [\sigma/2, \sigma].$$

Now let us suppose now that $\sigma \in (0, \frac{1}{2}]$ is given, take the unique integer $k \geq 1$ such that $\sigma \in [2^{-k-1}, 2^{-k})$, and assume that we have already established that

(12) $$\|\widetilde{u}(s) - \varphi\|_{L^2(S^{n-1})} < \eta, \qquad \forall\, s \in [\sigma, 1].$$

Then we can apply (10) with $1/2^\ell$ in place of σ for $\ell = 0, \ldots, k$, whereupon we obtain by summing over $\ell = 0, \ldots, j$

$$I(2^{-j})^{\alpha-1} - I(1)^{\alpha-1} \geq Cj, \qquad j \in \{1, \ldots, k\}.$$

For any $\tau \in [2^{-k}, \frac{1}{2}]$ there is an integer $2 \leq j \leq k$ such that $\tau \in [2^{-j}, 2^{-j+1}]$ hence this gives
(13)
$$I(\tau) \equiv \int_0^\tau r\|\frac{\partial\widetilde{u}}{\partial r}\|^2_{L^2(S^{n-1})}\, dr \leq \frac{C}{|\log\tau|^{1+\beta}}, \qquad \text{where } \beta = (1-\alpha)^{-1} - 1 > 0,$$

provided (12) holds. Of course since $I(\tau)$ is a decreasing function of τ this then also holds for $\tau \in [\sigma, 2^{-k}](\subset [2^{-k-1}, 2^{-k}])$ with a different (but still fixed) constant C and hence the formula (13) holds for all $\tau \in [\sigma, \frac{1}{2}]$, provided again that (12) holds. Notice integration by parts gives the general formula

$$\int_\sigma^1 |\log r|^{1+\beta/2} r f(r)\, dr = |\log r|^{1+\beta/2} \int_0^r sf(s)\, ds\Big|_\sigma^1$$
$$+ (1+\beta/2) \int_\sigma^1 r^{-1}|\log r|^{\beta/2} \int_0^r sf(s)\, ds\, dr$$

and using this with $f(s) = \|\frac{\partial\widetilde{u}(s)}{\partial r}\|^2_{L^2(S^{n-1})}$ we obtain by virtue of (13) that

(14) $$\int_\sigma^1 |\log r|^{1+\beta/2} r\|\frac{\partial\widetilde{u}}{\partial r}\|^2_{L^2(S^{n-1})}\, dr \leq C \int_\sigma^1 \frac{1}{s|\log s|^{1+\beta/2}} \leq C,$$

again subject to (12). But then we have by virtue of Cauchy-Schwarz that

$$\|\widetilde{u}(\sigma) - \widetilde{u}(\tau)\|_{L^2(S^{n-1})} \leq \int_\sigma^\tau \|\frac{\partial \widetilde{u}(r)}{\partial r}\|_{L^2(S^{n-1})}$$

(15)
$$\leq (\int_\sigma^\tau r|\log r|^{1+\beta/2}\|\frac{\partial \widetilde{u}(r)}{\partial r}\|^2)^{1/2}(\int_\sigma^\tau r^{-1}|\log r|^{-1-\beta/2})^{1/2}$$

$$\leq C|\log \tau|^{-\beta/4},$$

for any $0 < \sigma \leq \tau \leq \frac{1}{2}$, provided only that (12) holds. Next note that by direct application of the Cauchy-Schwarz inequality

$$\|\widetilde{u}(\tau) - \widetilde{u}(1)\|_{L^2(S^{n-1})} \leq \int_\tau^1 \|\frac{\partial \widetilde{u}}{\partial r}\|_{L^2(S^{n-1})}\,dr$$

(16)
$$\leq |\log \tau|^{1/2}(\int_{B_1} r^{2-n}|\frac{\partial \widetilde{u}}{\partial r}|^2)^{1/2} \equiv |\log \tau|^{1/2}\epsilon,$$

where $\epsilon = (\int_{B_1} r^{2-n}|\frac{\partial \widetilde{u}}{\partial r}|^2)^{1/2}$. By virtue of the triangle inequality, (16) alone guarantees that

(17)
$$\|\widetilde{u}(s) - \varphi\|_{L^2(S^{n-1})} < \eta/2, \qquad \forall\, \tau \leq s \leq 1$$

if

(18)
$$\epsilon|\log \tau|^{1/2} < \eta/4 \text{ and } \|\widetilde{u}(1) - \varphi\|_{L^2(S^{n-1})} < \eta/4.$$

So now suppose (18) holds with $\tau = \tau_0 \in (0, 1/2)$ such that $C|\log \tau_0|^{-\beta/4} < \eta/4$, and at the same time suppose that ϵ is chosen so that (18) holds with $\tau = \tau_0$. For any $\sigma \in (0, 1]$ we have by the triangle inequality that

(19) $\|\widetilde{u}(\sigma) - \varphi\|_{L^2} \leq \|\widetilde{u}(\sigma) - \widetilde{u}(\tau_0)\|_{L^2} + \|\widetilde{u}(\tau_0) - \widetilde{u}(1)\|_{L^2} + \|\widetilde{u}(1) - \varphi\|_{L^2}.$

(15), (16), (17), (18) (with $\tau = \tau_0$) and (19) we evidently deduce that

(20)
$$\|\widetilde{u}(s) - \varphi\|_{L^2(S^{n-1})} < 3\eta/4 \quad s \in [\sigma, 1];$$

all this was valid so long as (18) holds and our original assumption (12) holds; that is, assuming ϵ is small enough so that (18) (with $\tau = \tau_0$) holds, we have (12) $\Rightarrow$ (20) for any $\sigma \in (0, 1]$, which evidently establishes (20) for all $\sigma \in (0, 1]$, provided only we can ensure that ϵ can be selected so that (18) holds. However, $\widetilde{u} = u_{y,\rho}$, so by taking $\rho = \rho_j$ with j sufficiently large (where $\rho_j \downarrow 0$ is such that $u_{y,\rho_j} \to \varphi$) we can (and we do) ensure both inequalities in (18). (The second we already discussed at the beginning of the proof and the first trivially holds for j sufficiently large because $\int_{B_\rho} r^{2-n}|\frac{\partial u}{\partial r}|^2 \to 0$ as $\rho \downarrow 0$.) Thus we deduce that (20), and in particular (12), hold for all $\sigma \in (0, 1]$, and hence we can apply (15) with any σ, τ. Letting $\sigma = \sigma_j$ in (15) such that $\widetilde{u}(\sigma_j) \to \varphi$ in (15) (which we can do because φ is a tangent map of $\widetilde{u}$ at 0), we then have

$$\|\tilde{u}(\tau) - \varphi\|_{L^2(S^{n-1})} \leq C |\log \tau|^{-\beta/4}, \quad \tau \in (0, \tfrac{1}{2}],$$

which is the required asymptotic decay.

LECTURE 4
Recent Results on Rectifiability and Smoothness Properties of sing u

1. Statement of Main Theorems

We here want to begin by stating some of the recent theorems which have been proved about singularities of energy minimizing maps and minimal surfaces.

First recall that a subset $A \subset \mathbf{R}^n$ is said to be m-rectifiable if $\mathcal{H}^m(A) < \infty$, and if A has an approximate tangent space a.e. in the sense that for $\mathcal{H}^m$-a.e. $z \in A$ there is an m-dimensional subspace L_z such that

$$\lim_{\sigma \downarrow 0} \int_{\eta_{z,\sigma}(A)} f \, d\mathcal{H}^m = \int_{L_z} f \, d\mathcal{H}^m, \quad f \in C_c^0(\mathbf{R}^n),$$

where, here and subsequently, $\eta_{z,\sigma}(x) \equiv \sigma^{-1}(x-z)$ and $\mathcal{H}^m$ is m-dimensional Hausdorff measure. The above definition of m-rectifiability is well-known (see e.g. [SL4]) to be equivalent to the requirements that $\mathcal{H}^m(A) < \infty$ and that $\mathcal{H}^m$-almost all of A is contained in a countable union of embedded m-dimensional C^1-submanifolds of $\mathbf{R}^n$.

A subset $A \subset \mathbf{R}^n$ is said to be locally m-rectifiable if it is m-rectifiable in a neighbourhood of each of its points. Thus for each $z \in A$ there is a $\sigma > 0$ such that $A \cap \{x : |x-z| \leq \sigma\}$ is m-rectifiable. Similarly A is locally compact if for each $z \in A$ there is $\sigma > 0$ such that $A \cap \{x : |x-z| \leq \sigma\}$ is compact.

Then, with this terminology we have the following theorem from [SL8].

Theorem 1. *If u is an energy minimizing map of Ω into a compact real-analytic Riemannian manifold N, then, for each closed ball $B \subset \Omega$, $B \cap \mathrm{sing}\, u$ is the union of a finite pairwise disjoint collection of locally $(n-3)$-rectifiable locally compact subsets.*

Remarks: (1) Notice that being a finite union of locally m-rectifiable subsets is slightly weaker than being a (single) locally m-rectifiable subset, in that if $A = \cup_{k=1}^Q A_k$, where each A_k is locally m-rectifiable, there may be a set of points y of positive measure on one of the A_ℓ such that $\mathcal{H}^m((\cup_{k \neq \ell} A_k) \cap B_\sigma(y)) = \infty$ for each $\sigma > 0$. (This is possible because A_k has locally finite measure in a neighbourhood of each of its points, but may not have locally finite measure in a neighbourhood of points in the closure $\overline{A_k}$ and this may intersect A_ℓ, $\ell \neq k$.)

(2) It is also proved in [SL8] that $\Theta_u(z)$ is a.e. constant on each of the sets in the finite collection referred to in the above theorem, and that $\mathrm{sing}\, u$ has a (unique) tangent plane in the Hausdorff distance sense at $\mathcal{H}^m$-almost all points $z \in \mathrm{sing}\, u$, and u itself has a unique tangent map at $\mathcal{H}^m$-almost all points of $\mathrm{sing}\, u$. (See the discussion of Lecture 3 for terminology.)

There is an important refinement of Theorem 1 in case

(1) $\dim \operatorname{sing} u \le m$

for all energy minimizing maps into N.

In this case the conclusion of Theorem 1 holds with m in place of $n - 3$:

Theorem 2. *If u, N are as in Theorem 1, if $m \le n - 3$ is a non-negative integer, and if (1) holds, then for each closed ball $B \subset \Omega$, $B \cap \operatorname{sing} u$ is the union of a finite pairwise disjoint collection of locally m-rectifiable locally compact subsets.*

Remark. As for Theorem 1, again $\Theta_u(z)$ is constant a.e. on each of the sets in the finite collection referred to in the statement, $\operatorname{sing} u$ has a tangent space in the Hausdorff distance sense, and also u has a unique tangent map, at $\mathcal{H}^m$-almost all points of $\operatorname{sing} u$.

In [SL1,3] there are also results about singular sets (albeit for special classes of energy minimizing maps and stationary minimal surfaces), which, unlike the results here, were proved using "blowup methods". In particular we have

Theorem 3. *If $N = S^2$ with its standard metric, or if N is S^2 with a metric which is sufficiently close to the standard metric of S^2 in the C^3 sense, then $\operatorname{sing} u$ can be written as the disjoint union of a properly embedded $(n - 3)$-dimensional $C^{1,\mu}$-manifold and a closed set S with $\dim S \le n - 4$. If $n = 4$ then S is discrete and the $C^{1,\mu}$ curves making up the rest of the singular set have locally finite length in compact subsets of Ω.*

There are analogues of these theorems which apply to "multiplicity one" classes of stationary minimal surfaces.

For example, we have the following results:

Theorem 4. *(i) If M is the regular set of an n-dimensional mod 2 mass minimizing current in $\mathbf{R}^{n+k}$ (n, $k \ge 2$ arbitrary), then the singular set $\operatorname{sing} M$ is locally a finite union of locally $(n - 2)$-rectifiable, locally compact subsets.*
(ii) If V is an arbitrary n-dimensional stationary integral varifold in $\mathbf{R}^{n+k}$, if $0 \in \operatorname{sing} V$, and if the density of V at 0 is less than 2, then there is $\rho > 0$ such that $B_\rho(0) \cap \operatorname{sing} V$ is the union of an $(n-1)$-dimensional $C^{1,\alpha}$-manifold and a closed set of Hausdorff dimension $\le n - 4$.
(iii) If M is the regular set of an arbitrary n-dimensional mass minimizing current in $\mathbf{R}^{n+1}$, then $\operatorname{sing} M$ can locally be expressed as the finite union of locally $(n - 7)$-rectifiable, locally compact subsets.

Except for the local compactness result in part(i), parts(i) and (ii) of the above theorem are proved in [SL2]; the local compactness follows from the slight modification of the main argument of [SL2], following exactly the corresponding argument in §7 of [SL8]. This, and the proof of (iii) are described in detail in [SL9].

2. Brief Discussion of Techniques

We here want to outline a few of the main steps in the proof of the above theorems.

All of the results above concerning rectifiability depend on a rectifiability lemma, which we first want to describe. (The results of [SL2] also depend on such a rectifiability lemma, although in [SL2] the relevant result is not separated from the argument of the main proof; isolating the rectifiability lemma simplifies the presentation quite a bit.)

For the statement, we need to introduce a little more terminology.

Let $B_{\rho_0}(x_0)$ be an arbitrary ball in $\mathbf{R}^n$, and assume that $S \subset \overline{B}_{\rho_0}(x_0)$ is closed, that $\epsilon, \delta \in (0,1)$ with $\epsilon < \delta/8$ (in the applications below we always have $\epsilon << \delta$), and that S has the appropriate weak ϵ-approximation property. Specifically, for each $y \in S$ and each $\rho \in (0, \rho_0]$ we assume

$$(1) \qquad \begin{array}{l} S \cap B_\rho(y) \subset \text{ the } (\epsilon\rho)\text{-neighbourhood} \\ \qquad\qquad \text{of some } m\text{-dimensional affine space } L_{y,\rho} \text{ containing } y. \end{array}$$

(This is the same as the ρ_0-uniform weak ϵ-approximation property in the terminology of §3 of Lecture 2.)

In all that follows we assume that $L_{y,\rho}$, corresponding to each $y \in S$ and $\rho \le \rho_0$, is chosen. Then, relative to such a choice, we have the following definition.

Definition: With the notation in (1) above, we say S has a δ-gap in a ball $B_\rho(y)$ with $y \in S$ if there is $z \in L_{y,\rho} \cap \overline{B}_{(1-\delta)\rho}(y)$ such that $B_{\delta\rho}(z) \cap S = \emptyset$.

With this terminology and for any given $y \in S$, $\rho \in (0, \rho_0]$ and $\delta \in (0, \frac{1}{2})$, we define

$$\gamma(y, \rho, \delta) = \sup(\{0\} \cup \{\sigma \in (0, \rho] : S \text{ has a } \delta\text{-gap in } B_\sigma(y)\}).$$

Thus $\gamma(y, \rho, \delta) = 0$ means that S has no δ-gaps in the balls $B_\tau(y)$, $\tau \in (0, \rho]$, and if $\gamma(y, \rho, \delta) > 0$ then S has no δ-gaps in the balls $B_\tau(y)$, $\tau \in (\gamma(y, \rho, \delta), \rho]$, but S does have a δ-gap in $B_{\tau_j}(y)$ for some sequence $\tau_j \uparrow \gamma(y, \rho, \delta)$; in particular, S has a $\frac{\delta}{2}$-gap in $B_{\gamma(y,\rho,\delta)}(y)$ in this case.

Then the rectifiability lemma is as follows:

Lemma 1 (Rectifiability Lemma.). *Let $\delta \in (0, \frac{1}{32})$ be given. There is $\epsilon = \epsilon(m, n, \delta) \in (0, \frac{\delta}{16})$ such that the following holds. Let $\rho_0 > 0$, $x_0 \in S = a$ closed subset of $\mathbf{R}^n$ satisfying the ϵ-approximation property §2(1), and suppose:*
(I) Either S has a $\frac{\delta}{20}$-gap in $B_{\rho_0}(x_0)$ or there is an m-dimensional subspace $L_0 \subset \mathbf{R}^n$ and a family $\mathcal{F}_0$ of balls with centers in $S \cap \overline{B}_{\rho_0}(x_0)$ such that

$$(a) \qquad \sum_{B \in \mathcal{F}_0} (\operatorname{diam} B)^m \le \epsilon\rho_0^m,$$
$$(b) \qquad S \cap B_\sigma(y) \subset \text{ the } (\epsilon\sigma)\text{-neighbourhood of } y + L_0,$$
$$y \in S \cap B_{\frac{\rho_0}{2}}(x_0) \backslash (\cup \mathcal{F}_0), \ \sigma \in (\gamma(y, \frac{\rho_0}{2}, \delta), \frac{\rho_0}{2}]$$

(with $\gamma(y,\rho,\delta)$ as above), and

(II) $\forall x_1 \in S \cap \overline{B}_{\rho_0}(x_0)$ *and* $\forall \rho_1 \in (0, \frac{\rho_0}{2}]$ *there are* $L_1, \mathcal{F}_1$ *(depending on* x_1, ρ_1*) such that the hypotheses* (I) *continue to hold with* $x_1, \rho_1, L_1, \mathcal{F}_1$ *in place of* $x_0, \rho_0, L_0, \mathcal{F}_0$ *respectively. Then* $S \cap \overline{B}_{\rho_0}(x_0)$ *is* m-*rectifiable.*

Remarks: (1) It is important, from the point of view of the application which we have in mind, that the property I(b) need only be checked on balls $B_\rho(y)$ such that S has no δ-gap in any of the balls $B_\tau(y)$, $\rho \leq \tau \leq \rho_1/2$.

(2) Notice that if S does not have a $\frac{\delta}{20}$-gap in $B_{\rho_0}(x_0)$ (so that the first alternative hypothesis of the lemma does *not* hold), then, provided ϵ is sufficiently small relative to δ, no ball $B_\tau(y)$ for $\tau \in [\frac{\rho_1}{16}, \frac{\rho_1}{2}]$ and $y \in S \cap B_{\rho_0/2}(x_0)$ can have a δ-gap, so in particular condition I(b) always has non-trivial content in this case.

(3) The idea is to show (in the proof of Theorem 1 for example) that this lemma can be applied with $S = \overline{B}_\rho(y) \cap \{x \in \text{sing}\,u : \Theta_u(x) \geq \Theta_u(y)\}$ with $y \in \text{sing}\,u$ and with ρ sufficiently small. Notice that Lemma 3 of §4 of Lecture 2 already establishes the weak ϵ-approximation property (1) for such S, which is required before the above lemma can be used.

In the proof of the rectifiability lemma, we shall need the following covering lemma; for the proof of this, we refer to [SL8].

Lemma 2. *If* $\delta \in (0, \frac{1}{2})$, $F \subset \overline{B}_{\rho_0}(x_0) \subset \mathbf{R}^m$ *is arbitrary, and if* $\mathcal{B}$ *is a collection of closed balls of radius* $\leq \frac{\rho_0}{8}$ *and centers in* F *which covers* F, *and for each* $B = \overline{B}_\rho(y) \in \mathcal{B}$ *there is* $z \in \overline{B}_{(1-\delta)\rho}(y)$ *such that* $B_{\delta\rho}(z) \cap F = \emptyset$ *(that is, F has a δ-gap in each ball* $B \in \mathcal{B}$*), then there is a countable collection* $\mathcal{U} = \{B_{\rho_k}(y_k)\}$ *of balls with centers* $y_k \in F$ *which cover* F *and which satisfy*

$$\sum_k \rho_k^m \leq (1-\theta)\rho_0^m, \quad \theta = \theta(\delta, m) \in (0, 1),$$

and, also, for each $B_{\rho_k}(y_k) \in \mathcal{U}$ *there is a ball* $\overline{B}_{\sigma_k}(z_k) \in \mathcal{B}$ *such that* $B_{\theta^{-1}\rho_k}(y_k) \supset B_{\sigma_k}(z_k)$.

Proof of the Rectifiability Lemma: We assume first that S has no $\frac{\delta}{20}$-gap in $B_{\rho_0}(x_0)$, and hence (Cf. Remark (2) above), for $\epsilon \leq \epsilon_0$, with $\epsilon_0 = \epsilon_0(n, m, \delta)$ small enough,

(1) S *has no* δ-*gap in any ball* $B_\tau(y)$, $y \in \overline{B}_{\frac{\rho_0}{2}}(x_0) \cap S$, $\tau \in [\frac{\rho_0}{16}, \frac{\rho_0}{2}]$,

and also the hypotheses I(a), (b) must hold. Let

(2)
$$S^{(1)} = \{y \in S \cap \overline{B}_{\frac{\rho_0}{2}}(x_0) \backslash (\cup \mathcal{F}_0) : S \text{ has no } \delta\text{-gap in } B_\rho(y) \,\forall \rho \in (0, \tfrac{\rho_0}{2}]\},$$

and let

(3)
$$E_1 = S \cap \overline{B}_{\frac{\rho_0}{2}}(x_0) \backslash (S^{(1)} \cup (\cup \mathcal{F}_0)).$$

For each $y \in S^{(1)}$ we have by the property (b) that S satisfies the uniform cone condition

(4) $$S \cap B_\rho(y) \subset \text{ the } (\epsilon\rho)\text{-neighbourhood of } y + L_0$$

for each $\rho \in (0, \frac{\rho_0}{2}]$, and hence we deduce immediately that

(5) $$S^{(1)} \subset G_1, \quad \mathcal{H}^m(S^{(1)}) \leq \omega_m(\rho_0/2)^m,$$

where G_1 is the graph of a Lipschitz function f defined over the m-dimensional subspace L_0 with $\text{Lip} f \leq C\epsilon$. Let P_0 be the orthogonal projection of $\mathbf{R}^n$ onto L_0.

By (1) and property §2(1) we have that for each $y \in E_1$

(6) S has a $\frac{\delta}{2}$-gap in $B_{\rho_y}(y)$, S has no δ-gap in $B_\tau(y) \forall \tau \in (\rho_y, \frac{\rho_0}{2}]$,

where $\rho_y = \gamma(y, \frac{\rho_0}{2}, \delta) \in (0, \frac{\rho_0}{16}]$, and by hypothesis I(b)

(7) $$S \subset \{x \in B_{\rho_y}(y) : \text{dist}(x, y + L_0) \leq \epsilon\rho_y\} \cup K_{y,\epsilon,L_0},$$

where K_{y,ϵ,L_0} is the double cone $\{x \in \mathbf{R}^n : \text{dist}(x, y + L_0) \leq \epsilon|x - y|\}$.

Next, define $F = P_0(E_1)$, and let $\mathcal{B}$ be the collection of balls in L_0 which are orthogonal projections of the balls $\overline{B}_{\rho_y}(y)$, $y \in E_1$. For the remainder of this argument, balls in L_0 will be denoted $B_\rho^0(y)$. Thus $\mathcal{B} = \{\overline{B}_{\rho_y}^0(\tilde{y})\}_{y \in E_1}$, where $\tilde{y}$ is the orthogonal projection of y onto L_0. By virtue of (6), (7) we know that F has a $\frac{\delta}{4}$-gap in each of the balls of $\mathcal{B}$, and hence by the above covering lemma we have that there is a collection of balls $\tilde{\mathcal{B}} = \{\overline{B}_{\rho_k}^0(\tilde{y}_k)\}$, with $\tilde{y}_k = P_0(y_k)$, $y_k \in E_1$, which cover F and which satisfy $\sum_k \rho_k^m \leq (1 - \theta)(\rho_0/2)^m$, and for each k there is $\overline{B}_{\sigma_k}^0(\tilde{z}_k) = P_0(\overline{B}_{\sigma_k}(z_k)) \in \mathcal{B}$ (with $\sigma_k = \rho_{z_k}$), such that $B_{\sigma_k}^0(\tilde{z}_k) \subset B_{\theta^{-1}\rho_k}^0(\tilde{y}_k)$. By (7) with $y = z_k$ we have $S \subset \{x \in B_{\sigma_k}(z_k) : \text{dist}(x, z_k + L_0) \leq \epsilon\sigma_k\} \cup K_{z_k,\epsilon,L_0}$, and hence in particular $|y_k - z_k| \leq 2\theta^{-1}\rho_k$, $\text{dist}(y_k, z_k + L_0) \leq 2\theta^{-1}\epsilon\rho_k$, and

$$E_1 \cap P_0^{-1}(B_{\rho_k}^0(\tilde{y}_k)) \subset E_1 \cap B_{(1+6\theta^{-1}\epsilon)\rho_k}(y_k),$$

provided ϵ is sufficiently small relative to θ.

Thus, provided S has no $\frac{\delta}{20}$-gap in $B_{\rho_0}(x_0)$ and ϵ is sufficiently small relative to θ, we have constructed a countable collection of balls $\{B_{\tau_k}(y_k)\}$ ($\tau_k = (1 + 6\theta^{-1}\epsilon)\rho_k$) with centers in E_1 such that, after a change of notation (replacing θ by $2^{m+2}\theta$)

(8)
$$E_1 \subset \cup_k B_{\tau_k}(y_k), \quad \sum_k \tau_k^m \leq (1 - 2^{m+2}\theta)(\rho_0/2)^m + C\epsilon\rho_0^m \leq (1 - 2^{m+1}\theta)(\rho_0/2)^m$$

for suitable $\theta = \theta(n, m, \delta) \in (0, \frac{1}{2})$.

Now with $F = P(S \backslash B_{\rho_0/2}(x_0))$, P the orthogonal projection of $\mathbf{R}^n$ onto the affine space L_{x_0,ρ_0}, we can first cover all of $L_{x_0,\rho_0} \cap B_{\rho_0}(x_0) \backslash B_{\rho_0/2}(x_0)$

by balls $B_{\rho_k}(y_k)$ with centers y_k in L_{x_0, ρ_0} and radii $\rho_k \geq C^{-1}\rho_0$, with $C = C(m, \theta) > 0$, such that

$$(9) \qquad \sum_k \rho_k^m \leq (1 + \theta/2)(\rho_0^m - (\rho_0/2)^m).$$

But then (see e.g. Lemma 2.5 of [SL8] for the formal justification) using this collection $B_{\rho_k}(y_k)$ we can construct a new collection $\widetilde{\mathcal{B}}_0 = \{B_{\sigma_k}(z_k)\}$ with $z_k \in F$, $F \subset \cup \widetilde{\mathcal{B}}_0$, and such that $\sum_k \sigma_k^m \leq (1 + \theta)(\rho_0^m - (\rho_0/2)^m)$ and $\sigma_k \geq C^{-1}$, with $C = C(m, \theta)$. Using property §2(1) we can then use the $\widetilde{\mathcal{B}}_0$ to construct a collection $\mathcal{B}_1 = \{B_{(1+C\epsilon)\sigma_k}(\hat{z}_k)\}$ with $P(\hat{z}_k) = z_k$ and $\hat{z}_k \in S \backslash B_{\rho_0/2}(x_0)$ (so $|z_k - \hat{z}_k| < \epsilon\rho_0$ by 2.1) which covers all of $S \backslash B_{\rho_0/2}(x_0)$. Thus $\widetilde{\mathcal{B}} = \{B_{\tau_k}(y_k)\} \cup \mathcal{F}_0 \cup \mathcal{B}_1$ is a collection of balls with centers in $S \cap \overline{B}_{\rho_0}(x_0)$ with the the properties that

$$(10)$$

$$S \backslash S^{(1)} \subset \cup_{B \in \widetilde{\mathcal{B}}} B$$
$$\sum_{B \in \widetilde{\mathcal{B}}}(\operatorname{diam} B)^m \leq (1 - 2^{m+1}\theta)(\rho_0/2)^m + C\epsilon\rho_0^m + (1 + \theta)(\rho_0^m - (\rho_0/2)^m)$$
$$\leq (1 - \theta)\rho_0^m$$

(for ϵ sufficiently small depending on m, θ) and

$$(11) \qquad S^{(1)} \subset G_1 \cap \overline{B}_{\rho_0/2}(x_0), \quad \mathcal{H}^m(S^{(1)}) \leq \omega_m \rho_0^m,$$

where G_1 is the graph of a Lipschitz function f over L_0.

Of course if S does have a $\frac{\delta}{20}$-gap in the ball $B_{\rho_0}(x_0)$, then using §2(1) it is relatively trivial to find a cover $\widetilde{\mathcal{B}}$ of balls such that (10) and (11) hold with $S^{(1)} = \emptyset$. (For example one can use Lemma 2.5 of [SL8] to formally justify this.) Thus regardless of which of the alternative hypotheses of (I) hold, we always conclude that (10) and (11) hold.

We now proceed inductively. Assume that $J \geq 1$ and $S^{(j)} \subset S$, and that balls $\{B_{\rho_{j,k}}(x_{j,k})\}_{k=1,2,\ldots}$, $j = 1, \ldots, J$, are already constructed (with $x_{j,k} \in S$) so that $S^{(0)} = \emptyset$, $\{B_{\rho_{0,k}}(x_{0,k})\} = \{B_{\rho_0}(x_0)\}$,

$$S \backslash \cup_{j=1}^J S^{(j)} \subset \cup_k B_{\rho_{J,k}}(x_{J,k}),$$

$\cup_{j=1}^J S^{(j)}$ is contained in a countable union of Lipschitz graphs, and

$$\sum_k \rho_{j,k}^m \leq (1 - \theta) \sum_k \rho_{j-1,k}^m \text{ and } \mathcal{H}^m(S^{(j)}) \leq \omega_m \sum_k \rho_{j-1,k}^m, \quad j = 1, \ldots, J.$$

Then we repeat the argument described above, starting with $S \cap \overline{B}_{\rho_{J,k}}(x_{J,k})$ in place of S and with $x_{J,k}$, $\rho_{J,k}$ in place of x_0, ρ_0. Then conclusions (10), (11) imply that we have a Lipschitz graph Σ_k^J and a subset $S_k^{(J)} \subset S \cap \Sigma_k^J \cap \overline{B}_{\rho_{k,J}/2}(x_{k,J})$ such that $\mathcal{H}^m(S_k^{(J)}) \leq \omega_m \rho_{J,k}^m$, and balls $\{B_{\rho_{J,k,\ell}}(x_{J,k,\ell})\}_{\ell=1,2,\ldots}$ with centers in S such that

$$S \cap \overline{B}_{\rho_{J,k}}(x_{J,k}) \backslash S_k^{(J)} \subset \cup_\ell B_{\rho_{J,k,\ell}}(x_{J,k,\ell})$$

and

$$\sum_\ell \rho_{J,k,\ell}^m \le (1-\theta)\rho_{J,k}^m.$$

Relabelling so that $\{B_{\rho_{J,k,\ell}}(x_{J,k,\ell})\}_{k,\ell=1,2,\dots} = \{B_{\rho_{J+1,k}}(x_{J+1,k})\}_{k=1,2,\dots}$ and defining $S^{(J+1)} = \cup_k S_k^{(J)}$ we then have

$$\mathcal{H}^m(S^{(J+1)}) \le \sum_k \mathcal{H}^m(S_k^{(J)}) \le \omega_m \sum_k \rho_{J,k}^m$$

$$\sum_k \rho_{J+1,k}^m \le (1-\theta)\sum_k \rho_{J,k}^m,$$

and

$$S \backslash \cup_{j=1}^{J+1} S^{(j)} \subset \cup_k(S \cap B_{\rho_{J,k}}(x_{J,k}) \backslash S^{(J+1)}) \subset \cup_k B_{\rho_{J+1,k}}(x_{J+1,k}).$$

Thus such a collection exists for all J and

$$S \backslash \cup_{j=1}^J S^{(j)} \subset \cup_k B_{\rho_{J,k}}(x_{J,k}), \quad \sum_k \rho_{J,k}^m \le (1-\theta)^J \rho_0^m$$

$$\cup_{j=1}^J S^{(j)} \subset \text{ a countable union of Lipschitz graphs}$$

$$\mathcal{H}^m(\cup_{j=1}^J S^{(j)}) \le \sum_{j=1}^J (1-\theta)^{j-1} \omega_m \rho_0^m.$$

Thus $S \backslash (\cup_j S^{(j)})$ has $\mathcal{H}^m$-measure zero, $\cup_j S^{(j)}$ is contained in a countable union of Lipschitz graphs, and $\mathcal{H}^m(\cup_j S^{(j)}) \le C\rho_0^m$. Thus the lemma is proved.

The idea of the proof of Theorem 1 is to show that the above lemma can be applied with $n - 3 = m$ and with

$$S = \{x \in \text{sing } u \cap \overline{B}_{\rho_0}(y) : \Theta_u(x) \ge \Theta_u(y)\}$$

for a given $y \in \text{sing } u$ and for suitable $\rho_0 > 0$ (depending on u, N, y). Notice that we already checked in Lemma 3 of §4 of Lecture 3 that such an S does have the required weak ϵ-approximation property, so it remains to check the hypotheses I(a), (b) of the rectifiability lemma. This is too technical to describe here, but we do want to emphasize that the monotonicity identity §1(6) of Lecture 1, and variants of it, play a crucial role in developing the necessary energy and L^2 estimates which are needed before the hypotheses I(a), (b) can be checked. This is where we capitalize on the fact that hypothesis I(b) need only be checked in balls $B_\rho(y)$ such that no ball $B_\tau(y)$, $\tau \ge \rho$ has a δ-gap, because it turns out to be possible (using the monotonicity and its variants) to prove much stronger L^2 estimates for u in a ball where there are no δ-gaps.

References

[A1]   F. Almgren, *Q-valued functions minimizing Dirichlet's integral and the regularity of of area minimizing rectifiable currents up to codimension two,* Preprint.

[A2]   F. Almgren, *Existence and regularity almost everywhere of solutions to elliptic variational problems among surfaces of varying topological type and singularity structure,* Annals of Math. **87** (1968), 321–391.

[AW]   W. Allard, *On the first variation of a varifold,* Annals of Math. **95** (1972), 417–491.

[B]   F. Bethuel, *On the singular set of stationary harmonic maps,* Preprint

[BCL]   H. Brezis, J.-M. Coron, & E. Lieb, *Harmonic maps with defects,* Comm. Math. Physics **107** (1986), 82–100.

[DeG]   E. De Giorgi, *Frontiere orientate di misura minima,* Sem. Mat. Scuola Norm. Sup. Pisa (1961), 1–56.

[E]   C. L. Evans, *Partial regularity for stationary harmonic maps into spheres,* Arch. Rat. Mech. Anal., **116** (1991), 101–163.

[GG]   M. Giaquinta & E. Giusti, *The singular set of the minima of certain quadratic functionals,* Ann. Scuola Norm. Sup. Pisa **11** (1984), 45–55.

[GT]   D. Gilbarg & N. Trudinger, Elliptic Partial Differential Equations of Second Order, Springer-Verlag, 1983.

[GW]   R. Gulliver & B. White, *The rate of convergence of a harmonic map at a singular point,* Math. Annalen **283** (1989), 216–228.

[HF]   F. Hélein, *Régularité des applications faiblement harmoniques entre une surface et une variété Riemanninenne,* C.R. Acad. Sci. Paris **312** (1991), 591–596.

[HL1]   R. Hardt & F.-H. Lin, *The singular set of an energy minimizing harmonic map from B^4 to S^2,* Preprint 1990.

[HL2]   R. Hardt & F.-H. Lin, *Mappings minimizing the L^p norm of the gradient,* Comm. Pure & Appl. Math. **40** (1987), 555–588.

[JJ]   J. Jost, *Harmonic maps between Riemannian manifolds,* Proceedings of the Centre for Mathematical Analysis, Australian National University, **3** (1984).

[L]   S. Lojasiewicz, *Ensembles semi-analytiques,* IHES notes (1965).

[Lu1]   S. Luckhaus, *Partial Hölder continuity for minima of certain energies among maps into a Riemannian manifold,* Indiana Univ. Math. J. **37** (1988), 349–367.

[Lu2]   S. Luckhaus, *Convergence of Minimizers for the p-Dirichlet Integral,* Preprint, 1991.

[MCB]   C. B. Morrey, Multiple integrals in the calculus of variations, Springer Verlag, 1966.

[R]   R. E. Reifenberg, *Solution of the Plateau problem for m-dimensional surfaces of varying topological type,* Acta. Math. **104** (1960), 1–92.

[Riv]   E. Riviere, *Everywhere discontinuous maps into the sphere,* Preprint.

[SS]   R. Schoen & L. Simon, *Regularity of stable minimal hypersurfaces,* Comm. Pure Appl. Math. **34** (1981), 741–797.

[SU]   R. Schoen & K. Uhlenbeck, *A regularity theory for harmonic maps,* J. Diff. Geom. **17** (1982), 307–336.

[SL1]   L. Simon, Lectures on Geometric Measure Theory, Proceedings of the Centre for Mathematical Analysis, Australian National University, **3** (1983).

[SL2]   L. Simon, *Cylindrical tangent cones and the singular set of minimal sub-manifolds*, J. Diff. Geom. **38** (1993), 585–652.

[SL3]   L. Simon, *On the singularities of harmonic maps*, In preparation.

[SL4]   L. Simon, *Singularities of Geometric Varational Problems*, Summer School Lectures delivered at RGI, Park City, Utah, 1992, To appear in AMS Park City Geometry Series.

[SL5]   L. Simon, *Proof of the Basic Regularity Theorem for Harmonic Maps*, Summer School Lectures delivered at RGI, Park City, Utah, 1992, To appear in AMS Park City Geometry Series.

[SL6]   L. Simon, *Asymptotics for a class of non-linear evolution equations, with applications to geometric problems*, Annals of Math. **118** (1983), 525–572.

[SL7]   L. Simon, Lectures on regularity and singularity of energy minimizing maps: Monograph to appear in ETH Lectures in Mathematics series, Birkhäuser

[SL8]   L. Simon, *Rectifiability of the singular set of energy minimizing maps*, To appear in Calculus of Variations and PDE.

[SL9]   L. Simon, *Rectifiability of the singular set of multiplicity 1 minimal surfaces*, To appear in Surveys of Differential Geometry.

[WB]   B. White, *Non-unique tangent maps at isolated singularities of harmonic maps*, Bull. A.M.S. **26** (1992), 125–129.

Index